PT BOATS BEHIND THE SCENES

--ROBERT E. PICKETT born October 18, 1924, died December 8, 2010

FRANK J. ANDRUSS, SR.

NIMBLE BOOKS LLC

Nimble Books LLC

1521 Martha Avenue

Ann Arbor, MI, USA 48103

http://www.NimbleBooks.com

wfz@nimblebooks.com

+1.734-330-2593

Version 1.0; last saved 2011-01-21.

Printed in the United States of America

ISBN-13: 978-1-934840-85-6

Photo and Documentation Credits: National Archives, Andrew Shannahan, Electric Boat Company, Huckins Yacht Corporation, Alex Johnson, PT Boats Incorporated, Ed Behney, Navy War College Museum, R. Perry Collins, Cindy Purcell, Jerry E. Strahan, Bruce Marshall, Rick Grant , Capt. Robert J. Bulkley Jr., T. Garth Connelly, U.S. Naval Institute.

Contents

FOREWORD

Writing the foreword for this book is certainly an honor for me. It gives me a chance to talk about those men who served behind the scenes servicing the little PT boats of WWII.

I entered the Navy in late 1943 while working at Mare Island Naval Shipyard as an outside machinist. Skipping the chance for a rate, I had gone directly to boot camp at Farragut, Idaho, then to Diesel School at Iowa State. When I had finished Diesel School, my next appointment would be the Motor Torpedo Boat Squadrons Training Center, Melville, Rhode Island. School there was plenty tough, but I had made it through. When it came for me to ship out, I had a long trip to Treasure Island in California, and boarded the troop ship *General Mark L. Hersey* to Port Moresby, New Guinea. My first stop was Woendi Island and PT Base 21, where the Floating Equipment Maintenance Unit (FEMU) was being formed. We were to be sent where PT boat tenders could not get to, or were not needed.

Usually we ended up at small bases throughout the Philippines which had been retaken by our Allied forces. I was assigned to FEMU Crane Barge #6, which consisted of a 100-foot plus platform made with 4 by 4 by 4 foot steel pontoons. On our deck was a full size Quonset hut, used as a work shop, and living quarters for four men. This barge had two pusher units, using two Chrysler Marine Engines to move us when necessary, two 110 electric generators and a ten-ton crane. This crane was used to hoist fuel drums, torpedoes, engines, and other heavy items needed for the PT boats. The FEMU unit consisted of several crane barges; fuel barges, carrying 80,000 to 100,000 gallons of high-octane gasoline to fuel the boats; and dry docks. At any one time we had roughly 275 officers and men. The dry docks were instrumental in getting the boats out of the water for hull repair, painting, and propeller work. We also had an assortment of small craft (LCVPs and LCMs) which we used to haul parts, move barges, and personnel when needed. The barges and attending personnel traveled all over the Philippines. PT boats could not operate without the assistance of base forces. The base forces gave them refuge, a place to dock, repair facilities, a hospital, and food. The base forces also provided outdoor movies, special radio communications as well as their own base force quarters, for commanding officers and other non boat personnel assigned to each squadron. PT boats fought and operated in the Aleutians, Europe, Africa and the Pacific. Bases, tenders, and support groups were always there to assist them whenever they were needed.

It was certainly an interesting time for me during the war, one I shall never forget. We served with a great group of fellows that did their jobs day after day. The little PT boats were a wonderful machine in those days, one of the fastest on the water. They were capable of firing torpedoes, sinking ships, downing enemy aircraft, operating in shallow waters, and providing close support. They made history during the war as a small wooden boat, in more frequent contact with the enemy in close combat than any other surface craft. I was proud to be a part of the behind the scenes group that did so much to keep these little boats operating at full tilt to meet the enemy head on.

Alfred B. (Al) Moore

FEMU Crane Barge #6

PREFACE

When I had finished my first project, BUILDING THE PT BOATS, I thought about all of the work that went into the making of this book. I never pretended to be an author with that book, and I still feel the same way. Doing the book was my way of honoring those who worked so hard to complete these wonderful small boats and rush them to the war. In the back of my mind, I knew there were other areas concerning the PT boats that had not been touched on in recent years. These wonderful wooden boats that had captured the American people in the David vs. Goliath role during World War Two did not operate in the waters around the globe with just a trained crew. There was much more going on behind the scenes that people were just not aware of.

PT boats were an advanced technical piece of equipment for their time. Their high speed engines, weapons systems, and light weight frames needed constant attention and upkeep to keep them on the front lines. Special training was needed along with special ships and base force crews to do this massive job. As the war shifted to different locations, so too did the little PT boats, they needed a place that could handle these men and equipment. Once again my goal is to capture your imagination through actual photographs. You may notice some of the photos are not crisp and clear, but this is how they were shot at that time. Some sections in the book do not have many photos, so please if you do have any photographs related to the topics contained in this book, we would love to hear from you. I hope to honor those who worked so very hard, not as a crewmember of a fast sleek PT boat, but as a team player working behind the scenes to give those boys a fighting chance in a wonderful wooden boat.

Many of the men that made up these behind the scenes cast of characters were hard-working young men, who toiled day in and day out in the sweltering heat of the Pacific, the torpedo war of the Mediterranean, or the cold weather of the Aleutians. It was their job to see to it that these boats were kept ready to go, and the men well feed with a place to rest before the next patrol. Many times forward bases were nothing more than grass shacks and tents, but they continued to keep pace with the ever growing war effort. These bases and tenders were often in the front lines and under constant attack, but still the boats continued to be serviced. It is a testimony to the hard work and dedication of those who did their jobs to keep Uncle Sam's fighting PT boats always ready to meet the enemy head on.

ACKNOWLEDGMENTS

So many people were instrumental in helping me complete this book that I want to publically thank them. First, to my good friend Chip Marshall who always goes the extra mile. I appreciate your friendship and the hard work you did for me at the National Archives and at PT Boats Inc. To Alyce Guthrie, the mover and shaker of PT Boats Inc., your resources and kindness are so very much appreciated. I know your parents would be so proud of your efforts. To Alfred B. (Al) Moore, an original member of the FEMU barges, thanks for everything. To my friend, Robert Pickett, who painted the illustrations on the cover of this book and served on a PT boat during the war, I thank you. To my good friend Richard (Dick) Washichek, who through his kindness and dedication helped with many of the photographs in this book, providing me with CD's and photo sheets. To my friend Gene Kirkland, who helped me in Memphis Tennessee, I thank you. To my late and dear friend Wallace McNeish, who was there during the war, and served on board one of the wooden wonders of the sea. You were like my grandfather and I shall miss you very much. To my friend Harold E. Ted Walther Jr., who never refuses to help me in my continued quest for information concerning PT boats, I thank you, to my friend Jerry Gilmartin who is always there for me, to my wonderful wife Stacia, who continues to be my guiding light, my two sons, Frank Jr. and Tony, who I hope will learn to appreciate the history of the PT boats, and finally to all of those men who served behind the scenes to keep the crews and PT boats in fighting trim.

PHOTO AND DOCUMENTATION CREDITS

Captain Robert J. Bulkley Jr. USNR (retired); PT Boats Inc., Germantown, Tennessee; National Archives, Harold E. "Ted" Walther Jr.; Alyce Guthrie; Chip Marshall; Gene Kirkland; US Navy Department; Irwin Chase Sr.; Donald Millman; James Stanton; Earl Richmond; Alfred B. Moore; MTBSTC photo department; Arthur Frongello; Navy War College Museum; Robert Cembrola; Robert Douglas; Randy Finfrock; Will Day; At Close Quarters; Alvin Hansen; Randall McConnell; Wallace McNeish; C.J. Willis; United Press Photo; Darcy Kirk; Lt. Everett L. Carrier; Lt. Cmdr. Lester H. Gamble; John Kenney; US Naval Institute; Walter Pirog; Richard Washichek; Robert Pickett.

THE TENDERS

The first PT boat tenders were far from ideal, but they were the best available at the time. They were useful as communications centers, and could communicate with the boats on patrol and at the same time receive orders from and pass information to the higher echelons of command. They serviced torpedoes and performed a certain amount of engineering and electrical repair work. They could do some carpentry work, but unless they towed floating dry docks with them, could do little in the way of underwater repairs.

They carried gasoline, supplies, and equipment in limited quantities. They messed the boat crews aboard, sparing the always overworked PT generators the added burden of supplying power to operate the electric stoves. Sometimes when their evaporators were working well, they could offer the boat crews the luxury of fresh water showers.

The first tender was the *Niagara* (PG-52), the former Manville yacht *Hi-Esmaro*. She would be designated AGP-1. The *Niagara* was 235 feet long, 1,923 tons displacement. Built at the Bath Iron Works, Bath, Maine, in 1929, she was converted in New York in the winter of 1940. She was at Pearl Harbor for a time tending Squadron 1, then went to Panama, and underwent further conversion in New York in the summer of 1942. In the winter of 1942, she was assigned to the South Pacific area. On the morning of May 23 around 11:35, the *Niagara* and her boats from Squadron 23 were heading southwest towards Espiritu Santos, when a Japanese twin engine monoplane passed directly over the *Niagara*. The ship was called to general quarters and the PT boats were directed to make a wide circle around the tender.

The plane came back and dropped four bombs with all four being near misses. They were close enough however to disable the sound gear and the training mechanism of a 3 inch gun, and knock out the steering control temporarily. The *Niagara* opened fire with her guns, as six more enemy planes approached her one half hour after the first attack. The planes dropped at least 12 to 18 more bombs with one hit directly on the forecastle. The ship began listing to port as water rushed in from a 14-inch hole six feet below the waterline, flooding two storerooms and a passageway. Fires below deck started to become out of control and the engine room became flooded. The order was given to abandon ship as it was feared the fires would reach her gasoline storage tanks.

PT-146 and *PT-147* stood by to pick up those that went over the side, while others slide into rafts and were picked up by other PT boats. Despite her heavy damage, the *Niagara* would not sink. Even more important, not one of her 136 officers and men were killed or seriously injured. Because of limited fuel and the distance to Tulagi—230 miles—the PT boats had to head back without delay. The *Niagara* was doomed and it was decided to sink her. A torpedo from *PT-147* was fired, which struck the *Niagara* in the gasoline tanks. The ship exploded in a sheet of flame some 300 feet in the air, and went down in one minute. The *Niagara* was the only PT boat tender destroyed during the war, and earned one WWII battle star.

Because of confusion during the early stages of WWII, only vague specifications were received regarding what was required in a ship to make it suitable as a tender. Many of those who were on the early tenders had ever seen a PT boat, and to make matters even tougher, no one from the 11th Navy District knew the first thing about PT boats. The Navy was asked to send a qualified officer and as a result Lt. (jg) Henry M.S. Swift, USNR, was detached from Squadron 1 at Pearl Harbor and ordered to temporary duty with the second PT tender AGP-2 Hilo. This tender would be the last of the yachts used. His knowledge and experience were invaluable and responsible for what facilities the ship finally had. The tender *Hilo* would become the first tender in the Southwest Pacific area.

An important step in providing better tenders came with the commissioning of the USS *Portunus* in June 1943. This was a converted LST, 328 feet long, displacing 3,754 tons. The so called "tank deck," used in LSTs for cargo carrying space, was converted to shop space. Here at last was a PT tender with adequate space and adequate machinery. One of her best features was an "A-frame", a heavy device on the port side that could lift a PT boat out of the water for repairs to the underwater structure. The LST type ship was so satisfactory that eight more were built. Some of the later tenders were equipped with two A-frames, one on each side, so that two boats could be raised from the water simultaneously. Beginning with the *Oyster Bay,* commissioned in November 1943, four ships originally laid down as seaplane tenders were completed as PT tenders by their builder, the Lake Washington Shipyards, Houghton, Washington.

These ships were 310 feet long, and roughly 2,800 tons. They were fine, sleek ships, built along destroyer lines, and each carried, in addition to antiaircraft batteries, two 5 inch guns. Though they were faster than the slow LST type, they had limited shop space and had no means of raising a PT boat from the water unless they towed a dry dock. In certain types of operations, however, where speed and firepower were needed, they proved superior to the LST type.

Finally came two huge tenders, the *Acontius* (*AGP-12*) and *Cyrene* (*AGP-13*), commissioned in June and September of 1944. These ships were 413 feet long, with a full load displacement of 11,000 tons from converted Marine Commission C-1 hulls. These were splendid ships, with extensive shops and a vast amount of storage space. There was a boom capable of lifting a PT boat from the water and setting it down on the forecastle. The *Cyrene* obtained an extra cradle, so that it could set one boat down on the forecastle and lift another alongside at the same time. The only disadvantage of these ships was that they were too big and valuable to risk in an area where concentrated enemy air attack was likely-but they were ideal in the rear area.

Figure 1. A beautiful look at the graceful lines of the first PT boat tender, USS *Niagara (PG-52).* She was the former Manville yacht *Hi-Esmaro* built at the Bath Iron Works, Bath, Maine. She would later be designated *AGP-1,* serving for a time in Pearl Harbor tending Squadron 1. Assigned to the South Pacific area in the winter of 1942, she was jumped by Japanese planes, which dropped no less then eighteen bombs, with one hitting her forecastle. With the ship starting to list, she was abandoned, and later sunk by a torpedo from *PT-147.* This photo was taken from the deck of *PT-18* (Squadron 2) when the ship was in Cienfuegos, Cuba. (Irwin Chase Sr.) [1]

[1] The ship histories in this section draw heavily on the *Dictionary of American Naval Fighting Ships (DANFS),* available online at http://www.history.navy.mil/danfs/index.html.

Figure 2. *PT-18* (left) and *PT-11* are tied up to their tender, USS *Niagara*. *PT-11* had hit a shoal, causing damage to her hull. She was in danger of sinking and was towed to the tender in attempt to save her. In this photo, much of the equipment topside has been removed and hauled over to the *Niagara* to lighten the boat. Most of the work was handled by the crews of Squadron 2, although tender personnel did help out. Elco electrical engineer John C. Kenney, who was on the boat when it ran aground, managed to dine on the tender that night. These boats were seventy-foot designs from the Elco Naval Division and were attached to Squadron 2. During the time of this accident, the Squadron was operating off the Southern coast of central Cuba in Cienfuegos. (John C. Kenney)

Figure 3. A look at the harbor crane in Cienfuegos, Cuba on March 7, 1941 as this new 4M-2500 Packard Marine Engine is being installed on *PT-18* from Squadron 2. These were the first Elco-assembled boats and were tested in Florida and Caribbean waters. None of these boats would see action in the US Navy, and were lend-leased to the British Navy between April and July 1941. Tender personnel from the PT boat tender USS *Niagara* would assist with the engine change. (John C. Kenney)

Figure 4. A close-up look at the Packard Marine Engine being installed on *PT-18*. This massive power plant provided the much-needed power to bring about the high speed these little boats achieved. (John C. Kenney)

Figure 5. A look at some of the men assigned to the USS *Niagara (PG-52)* in 1940. These tender crews helped support Squadron 1 in Miami, which at that time consisted of experimental boats. At the time this photo was taken, the *Niagara* was in Cienfuegos, Cuba tending the first Elco seventy-foot PT boats of Squadron 2. The tender had limited repair facilities, but she was the best available at that time. The trip to Cuba and back proved that these little boats could withstand punishment, as they encountered extremely heavy seas but incurred little damage. (John C. Kenney)

Figure 6. USS *Hilo (AGP-2)*--Built as the yacht *Caroline* in 1931, this tender was purchased by the Navy in 1941. She was designated *PG-58* and commissioned as the *Hilo* on June 11, 1942. She became the first tender in the Southwest Pacific area, setting up the first PT boat base in New Guinea. She saw much action in Kana Kopa, undergoing many air raids. Shifting to Mios Woendi, she became the command ship for PT operations with her tender equipment removed. The Philippine invasion put her under constant air attack, her gunners scoring several kills. The *Hilo* received eleven battle stars, being decommissioned on March 3, 1946. (National Archives)

Figure 7. Another look at the PT tender USS *Hilo* as she tends her boats at Kana Kope, New Guinea. This was the first PT boat base set up in New Guinea, where the ship survived many Japanese air attacks. Notice the boats to the right of the photo are carrying the very early style SCR-517 Beehive radar domes. (Author's collection)

Figure 8. USS *Jamestown (AGP-3*). Built as the yacht *Savarona* in 1928, and considered the largest, most luxurious in the world at that time. She was purchased by the Navy in 1940, and converted to a gunboat *(PG-55),* renamed in May,1941. She was used to train Naval Academy midshipmen before being converted to a tender. She assisted with the forming of the Motor Torpedo Boats Squadron Training Center in the summer of 1941, before escorting convoys between New Hebrides and Tulagi. She towed barges loaded with gasoline and bombs, and helped repair cruisers. She received the Presidential Unit Citation for actions in Tulagi, Gavutu, Tanambogo, Florida, and Guadalcanal. She was decommissioned on March 6, 1945. (US Naval Institute)

Figure 9. October 11, 1940 shows the USS *Jamestown (PG-55)* at the Brooklyn Navy Yard in New York. The luxurious yacht *Alder* was purchased by the Navy on December 6, 1940 and converted into a gunboat in the Fletcher Division Shipyard of Bethlehem Shipbuilding Co., Hoboken, New Jersey. She was renamed *Jamestown* and designated *PG-55,* commissioned at New York Navy yard on May 26, 1941, Cmdr. A.P. Lawton in command. Here two seventy-seven-foot Elco PT boats from Squadron 2 are tied up next to her. In the lower right , *PT-37* while in command of J. J. Kelly was destroyed by Japanese destroyers on February 1, 1943 in the Savo-Esperance Channel area of Guadalcanal. Only one man would survive. (Harold E. Ted Walther Jr.)

Figure 10. Another nice look at the tender USS *Jamestown* shortly after her refit. Taken at the Navy yard in New York on July 23, 1942, the ship now carries a new paint scheme and better defensive armament. (Author's collection)

Figure 11. A look inside the torpedo shop on the PT boat tender USS *Jamestown.* Overhead hoists are made ready to haul out this 21 inch Mark VIII torpedo. They will be hauled up to the top deck, where they will be hoisted out over a waiting PT boat, and gently placed into the boat's tubes. Many times, the torpedoes were simply transferred from the ship, and brought to the shore for installation. A good supply of grease was brushed onto the tubes for easy access into the tubes and to prevent rust from exposure to salt water and humid climate. (PT Boats Inc., Germantown, Tennessee)

Figure 12. USS *Portunus (AGP-4)*. This was the first of the LST-type tenders, laid down as *LST-330* on November 12, 1942. Commissioned as *AGP-4* on June 12, 1943, she provided adequate space, machinery and an "A-frame", a heavy device to lift a PT boat from the water. Living space was limited onboard, so a 20 x 40 Quonset hut was erected on the forward deck. She serviced US and Australian naval units along the New Guinea coast, also supporting Squadrons 25, 31, 33, and 37. In 1945, she was convoyed to Okinawa and was decommissioned at Mare Island April 18, 1945. The tender earned three WWII Battle stars during her service. (National Archives)

Figure 13. This Squadron 25 boat has been placed into the A Frame sling. Some PT tenders had the ability to lift the boats out of the water, thus providing a way for visual inspection of the hull, below the water line. This was extremely important when repairs to the hull, running gear, or painting were needed. (Jerry Gilmartin)

Figure 14. USS *Varuna (AGP-5).* She was laid down on August 23, 1942 at Neville Island, Pa., by the Dravo Shipbuilding Yard. The ship was towed to Tampa, Florida where she was outfitted as a PT boat tender, and commissioned on August 31, 1943. She proceeded to Noumea, New Caledonia, and Espiritu Santo, New Hebrides, then to the Solomon Islands. Her first PT boat repair job was *PT-105* at Blanche harbor. Staying in the area, she repaired four boats per day before making her way to Green Island. The tender would also operate with Squadrons 27 and 28 at Mios Woendi on the northern tip of New Guinea. The *Varuna* was one of the first Allied warships to enter Manila Bay after the surrender of Corregidor. She proceeded to Bobon Point, Samar to assist with the decommissioning of the PT boats at Base-17. *Varuna* was decommissioned on January 4, 1946. She received four battle stars during WWII. (National Archives)

Figure 15. Hard-working men need three meals a day and these men are standing in the chow line on the PT boat tender USS *Varuna (AGP-5).* The ship was stationed in Blanche Harbor in the Treasury Islands when this photo was taken. The ship remained there for at least a month, repairing an average of four PT boats per day, before shifting her operations to Nissan Harbor, Green Island. (PT Boats Inc., Germantown, Tennessee)

Figure 16. A look at the machine shop located on the *Varuna.* These trained men are providing important metal parts that are vital to the PT boats' upkeep. In a very hot and humid place to work, these men kept up the pace so that needed repairs and new parts were always on hand. These are metal lathes and drill presses that required skilled workers. (PT Boats Inc., Germantown, Tennessee)

Figure 17. USS *Oyster Bay (AGP-6)* was laid down as *AVP-28,* a seaplane tender on, April 17, 1942. Reclassified as a PT boat tender, she was designed along the lines of a destroyer, faster than the LST type, but had limited shop space and no way to lift a PT boat from the water. If this was to be done, the tender would need to tow a drydock. She serviced two squadrons at Milne Bay, and then escorted fifteen boats to Seeadler Harbor, Admiralty Islands. The tender bombarded Pityili Island, and evacuated forty-two wounded to Finschhafen. She received heavy air attacks while en route to Leyte. While in Wasile Bay, the tender shot down one Japanese plane and assisted PTs and other tenders with knocking down at least three more attackers. She was decommissioned on March 26, 1946, but returned to the Navy in 1949, as *AVP-28.* She remained in the Pacific Reserve Fleet until being transferred to the Italian Government in 1957. She received five WWII battle stars. (PT Boats Inc., Germantown, Tennessee)

Figure 18. Another look at the tender *Oyster Bay* as she tends to her boats in Seeadler Harbor. The boats have taken up position alongside, with some pitching tents on their bows to keep crews out of the hot midday sun. Although she had the graceful lines of a destroyer, she had limited space onboard with no way to lift a PT boat from the water. (PT Boats Inc., Germantown, Tennessee)

Figure 19. USS *Mobjack (AGP-7).* Also laid down as a seaplane tender *(AVP-27),* she was reclassified as a PT boat tender on March 11, 1943, and began duty with MTBRons, SoPac. She arrived at Dreger Harbor on July 27, 1944 for service under ComMTBRons, 7th Fleet. *Mobjack* tended Squadron 33 and provided assistance with salvage. Sailing for Morotai, she would work with her squadron, also doing work with PBYs. *Mobjack* also transported materials, spares, and advance base personnel of Squadrons 9 and 10. She relieved the tender *Willoughby,* working with Squadrons 20 and 23. Her last stop was helping with the decommissioning of the PT boats in the Philippines, before being decommissioned and transferred to Department of Commerce, August 21, 1946. She later joined the US Coast and Geodetic Survey as the survey ship, *Pioneer.* During WWII she earned three battle stars. (PT Boats Inc., Germantown, Tennessee)

Figure 20. USS *Wachapreague(AGP-8).* Laid down as seaplane tender *AVP-56,* and commissioned on May 17, 1944. She made brief stops at Espiritu Santo, New Hebrides, and Brisbane, Australia, before reaching Milne Bay, New Guinea, on August 20, 1944. She dropped anchor at Base 21, tending to the PT boats of Squadron 12. Nightly operations took a heavy toll on Japanese barge traffic, and wrecked havoc on Japanese shore installations. As Allied forces wrapped up the New Guinea campaign, *Wachapreague* received an additional five PT boats from Squadron 7. The Navy was now preparing for operations to liberate the Philippines from Japanese occupation. On October 13, 1944, the tender sailed for Leyte, a 1,200 mile trek across open waters. With her were two other PT boat tenders, *Oyster Bay and Willoughby,* not to mention 45 PT boats. Along the way the small PT boats were fueled two at a time by the *Wachapreague,* until her fifteen boats had their fuel supply replenished. All the boats made this journey under their own power. The tender participated in the Battle of Surigao Strait. The boats met the heavy ships of the the Japanese Southern forces head on and their presence helped distract them. The tender worked with Squadrons 13, 16, and 28, as well as Squadrons 17 and 36 in San Pedro Bay. She was decommissioned May 10, 1946 and transferred to the United States Coast Guard on May 27, 1946. She earned four battle stars during WWII. (National Archives)

Figure 21. USS *Willoughby (AGP-9).* Laid down as the seaplane tender designated *AVP-57* on March 15, 1943, at Houghton, Washington. She was later reclassified as a PT boat tender on May 11, 1943. She left California bound for Funafuti in the Ellice Islands, and while enroute had her destination changed to Manus Island in the Admiralty Islands. She fueled and took on stores at Tulagi in the Solomon Islands before reaching her destination, Seeadler Harbor, Manus, on September 6, 1944. *Willoughby* tended boats at Mios Woendi for a short period before making a 1,200 mile journey to Leyte Gulf in the Philippines. She steamed to San Pedro Bay, tending boats from Squadrons 7 and 21 over the ensuing weeks. Japanese air attacks picked up, and at one point the *Willoughby* was at general quarters for over eleven hours. She shot down a Japanese plane during this engagement.. While at San Pedro Bay, she experienced a storm of typhoon intensity on October 30, 1944, but suffered no damage. From October 31st to November 12, 1944, she experienced daily air raids, and saw *PT-320,* moored off the tender's starboard bow, receive a direct hit, killing nearly the entire crew. *Willoughby* later steamed to Mios Woendi where she loaded stores and supplies for the long trip back to the Philippines. She later carried out extensive surrender operations in the vicinity of Kuching, Borneo, where she evacuated 210 Allied POWs and internees from Kuching to Labuan Islands. The tender would later serve in the Viet Nam War, under the name USCGC *Gresham.* She was decommissioned on April 25, 1973. She earned three battle stars during WWII. (National Archives)

Figure 22. *Willoughby* anchors off the mouth of the Sarawak River, near Tanjong Po, Borneo on September 1945. With her are *PT-224* (left) and *PT-299.* They are two of six boats from Squadron 16 that participated in the evacuation of hundreds of internees from Kuching Borneo. Here, the boats are crowded with men from the 9th Australian Division, who were part of the reoccupation troops, who were taken to the Sarawak capital. (PT Boats Inc., Germantown, Tennessee)

Figure 23. USS *Orestes (AGP-10)*. Laid down as *LST-135,* she was converted into a PT boat tender at the Maryland Drydock Company. Commissioned on April 25, 1944, she made her way to New Guinea, beginning operations at Aitape on August 23, 1944. The tender would make the move one month later, sailing to Mios Woendi, and then joining General Douglas MacArthur's invasion forces at Leyte. There, the gunners of the tender claimed their first Japanese kills of the war, two Mitsubishi A6M "Zeke" (Zero) fighters. While *Orestes* was en route to Mindoro with 30 PT boats and 50 other vessels, Japanese planes made life tenuous. On December 30, 1944 a Japanese dive bomber came in low on the starboard side and crashed into *Orestes* amidships, causing heavy damage and killing 45 members of her crew. After finally bringing the fires under control, *Orestes* was beached. Shortly afterwards, USS *LST-708* towed her back to Leyte for temporary repairs. She left Leyte on February 24, 1945 on a slow voyage back to the United States and arrived at San Francisco, California on May 13, 1945. She was completely repaired and set out for a second trip to the Pacific war zone, but the war end before she reached Guinan Harbor, Samar in the Philippines. She served with the Philippine Sea Frontier until December 17, 1945. She was decommissioned on April 29, 1946, earning two battle stars for her service in WWII. (PT Boats Inc., Germantown, Tennessee)

Figure 24. This photo shows the destruction to the PT boat tender *USS Orestes.* Notice the heavy damage caused when a Japanese Dive bomber crashed onto the deck, killing 45 crew members. She would be repaired, but arrive back in the Pacific after the war had ended. (PT Boats Inc., Germantown, Tennessee)

Figure 25. USS *Silenus (AGP-11).* Laid down as *LST-604,* she completed her shakedown cruise before departing for the Pacific Fleet. Her orders were to proceed to Tulagi, British Solomon Islands, and she arrived on October 27, 1944. She tended the PT boats of Squadron 37 until December, when she sailed for the Treasury Islands. She remained there for six weeks and then sailed for Espiritu Santo via Tulagi and San Cristobal. She remained there for almost six months, tending Squadrons 32 and 37. She later sailed to Okinawa, where her decommissioning was planned, but instead she was routed to New York, via Samar, P.I., Guam, and Pearl Harbor. She arrived in New York on January 17, 1946 and was decommissioned on March 14, 1946. *Silenus* would earn one battle star for her service in WWII. (Author's collection)

Figure 26. *USS Acontius (AGP-12).* A C1-A type tender, originally named *Cape Carthage,* she was one of the best-equipped tenders. Commissioned in June of 1944, she had extensive shops and a vast amount of storage space. She carried a boom that was capable of picking up a PT boat out of the water and setting it on the forecastle on a cradle. The disadvantage was that she was too big and valuable to risk in concentrated enemy attack areas; but she was ideal in the rear area. Following her shakedown out of Guantanamo Bay, Cuba, she steamed for Tulagi in the Pacific. She arrived in the Solomons on September 18, and except for a brief run to the Russell Islands in October, she remained anchored in Tulagi, tending the PT boats of Squadron 31. On November 3rd, she moved to Blanche Harbor, Treasury Islands, resumed her tender duties. Moving once again in December, *Acontius* got underway for the Palau Islands in company with Squadron 30 and 31, and stayed in that area until March 28, 1945. Once again the tender raised anchor, heading for the Philippines and dropping anchor at San Pedro Bay, where she provided services to several Squadrons. In August, she began serving as the flagship for Commander, Motor Torpedo Boat Squadrons, Pacific Fleet. After Japan surrendered, the *Acontius* participated in the inactivation of the PT boats at Base 17 in Samar. She was decommissioned on March 22, 1946. (National Archives)

Figure 27. USS *Cyrene (AGP-13).* Another C1-A type hull, this tender was laid down as *Cape Farewell,* under contract at Pusey and Jones Corporation, Wilmington, Delaware. Commissioned on September 27, 1944, she made her way to Manus Islands, arriving on December 13th to escort two squadrons of PT boats to Hollandia, New Guinea. This tender had three booms, one to lift a PT onto the deck, one to hang a PT at deck level, and one to lift onto a Seabee pontoon barge alongside. At 413 feet long, she had major repair facilities and berthing space, including large areas for storage and supplies. She sailed on convoy duty to Leyte, arriving on January 1, 1945. The *Acontius* served as flagship for Commander, Motor Torpedo Boat Squadrons, 7th Fleet. She sailed from Samar in the Philippines, arriving in the United States on January 7, 1946, reporting to the 12th Naval District for repair work in decommissioning small craft. She was decommissioned on July 2, 1946. Comedian and actor Don Rickles served as S1/c aboard this tender until 1946. (PT Boats Inc., Germantown, Tennessee)

Figure 28. USS *Alecto (AGP-14).* Laid down as *LST-977* on December 12, 1944. She was commissioned as a tender on February 8, 1945, and after her shakedown tended PT boats at the Motor Torpedo Boat Squadron Training Center in Melville. She then made her way to the Regional Development Center, Solomons, MD. At 328 feet long, this tender had plenty of room on deck for repairs, and a forward boom for lifting a PT boat from the water. She was be decommissioned on July 28, 1945. (PT Boats Inc., Germantown, Tennessee)

Figure 29. Another look at *Alecto* as she lifts this PT boat from the water. This photo was taken in Albany, New York during her shakedown period. She is lifting PT-505 from the water. The PT boat would first serve with Squadron 34, before being returned to the United States, serving as a training boat with Squadron 4 at the Motor Torpedo Boat Squadrons Training Center. (PT Boats Inc., Germantown, Tennessee)

Figure 30. USS *Callisto (AGP-15).* Laid down as *LST-966* on October 31, 1944, she was converted to a PT boat tender and commissioned on June 12, 1945. She sailed from Yorktown, Virginia, on July 23, 1945 bound for the Pacific and service with the 7th fleet. Delayed at Pearl Harbor by the cessation of hostilities, she reported to San Pedro Bay, Philippine Islands, on October 15th to serve as tender to Squadron 9. Callisto provided berthing, maintenance, and supply facilities for the squadron until December 20, 1945. She sailed for San Francisco and was decommissioned on May 9, 1946. (PT Boats Inc., Germantown, Tennessee)

Figure 31. USS *Antigone II (AGP-16).* Former *LST-733* was commissioned in reduced status on November 17, 1944. She reached Okinawa after the war ended. After her conversion to a PT boat tender at the Maryland Drydock Company, she joined the Pacific fleet in July of 1945. While at Okinawa, she provided service to Squadrons 31, 32, and 37. Leaving for San Francisco in December, this large tender arrived on May 27, 1946 where she was decommissioned and scrapped. (PT Boats Inc., Germantown, Tennessee)

Figure 32. USS *Brontes (AGP-17).* Although classified as a tender on August 14, 1944, she would be launched on February 6, 1945 as *LST-1125* by Chicago Bridge and Iron Company, and placed in reduced commission on March 10, 1945. After undergoing conversion to a PT boat tender, she was recommissioned on September 14, 1945. She sailed for New Orleans on September 26, 1945, arriving there on October 3rd. She participated in Navy Day activities and then remained to service the PT boats. In December 1945, she sailed for Washington, D.C., to participate in the "parade of torpedo boats" held in conjunction with a Victory Bond Drive. She was decommissioned in New York on March 14, 1946. Later she was converted as a merchant ship, serving under this capacity until she was wrecked by a hurricane on October 27, 1959. (PT Boats Inc., Germantown, Tennessee)

Figure 33. USS *Chiron (AGP-18).* Launched as *LST-1133,* she was converted and placed in reduced commission. After her conversion, she sailed for Miami, Florida where she serviced Squadron 42. *Chiron* remained in Miami until December, where she would sail for New York and was decommissioned on February 20, 1946. She was refitted as a merchant ship, performing these duties until she was wrecked on Manuel Luis Reef in 1960. (PT Boats Inc., Germantown, Tennessee)

Figure 34. USS *Pontus (AGP-20).* Launched as *LST-201* on March 2, 1943 in reduced commission. She was ferried down the Mississippi River to Algiers, LA. After her shakedown in Florida, she sailed with *LCT-254* on her deck for the Pacific, arriving in Brisbane on August 11th. She continued her conversion in Mackay with the installation of water distillers, machine and carpentry shops, and extra generator, a ten-ton crane, and embarkation of a Navy repair crews. She arrived in Milne Bay on October 18th and finished conversion, continuing on to Dreger Harbor. *Pontus* tended Squadrons 7 and 8 from Celco Island, off Aitape. It was not until August 15, 1944 that she was renamed Pontus, after overhaul in Brisbane. She would return to Mios Woendi on November 17. After the war ended, she performed tending and decommission duties at Subic Bay and Guiunan Harbor. She was decommissioned on April 2, 1946, and earned three battle stars during WWII. (PT Boats Inc., Germantown, Tennessee)

Figure 35. A look at the PT tender *USS Tatoosh (YAG-1).* While Base 5 at Finger Bay, Adak, Alaska was being constructed, this ship served as a tender. The ship had limited facilities, but it did have good food, hot showers, and motion pictures every night. It served Squadrons 13 and 16 in 1943. Although not officially a tender, it was used as a hospital ship. (Author's collection)

Figure 36. A nice shot of this PT boat as she approaches the USS *Gallup (PF-47),* a Patrol Frigate. Most likely she is looking for a quick hand out of food or other supplies that were in short supply for the PT fleet. Many surface combatants served as what we would call today "virtual tenders" for the scrappy PT fleet. This photo was taken in the Biak Island area of New Guinea. The frigate was built for anti-submarine warfare and operated with Escort Division 43. She bombarded enemy strongholds in Biak to support the landings at Blue Beach, north of Wardo River. (Author's collection)

Figure 37. *PT-160* from Squadron 9 undergoes refueling. Notice this PT boat tender has lowered her boom so that the boat could be pulled along. The boats engines would be shut down and lines attached to the boom while the tender was underway. Notice the home-made tarp that was attached to the helm area of the boat, and the three fueling lines for each of the boat's three gasoline tanks deep in the boat. (Alvin Hansen)

Figure 38. *PT-573* has been secured in the cradle and takes a short ride on the lift crane. Tender crews can now inspect the boat's hull for any damage and make necessary repairs. Having this type of crane lift greatly expanded the tender's role during the Pacific campaign and made things much easier for those working on the hull of the boats. This LST-type tender had a vast amount of shop space onboard, and another lift capable of lifting the boat and setting it down on deck. (Author's collection)

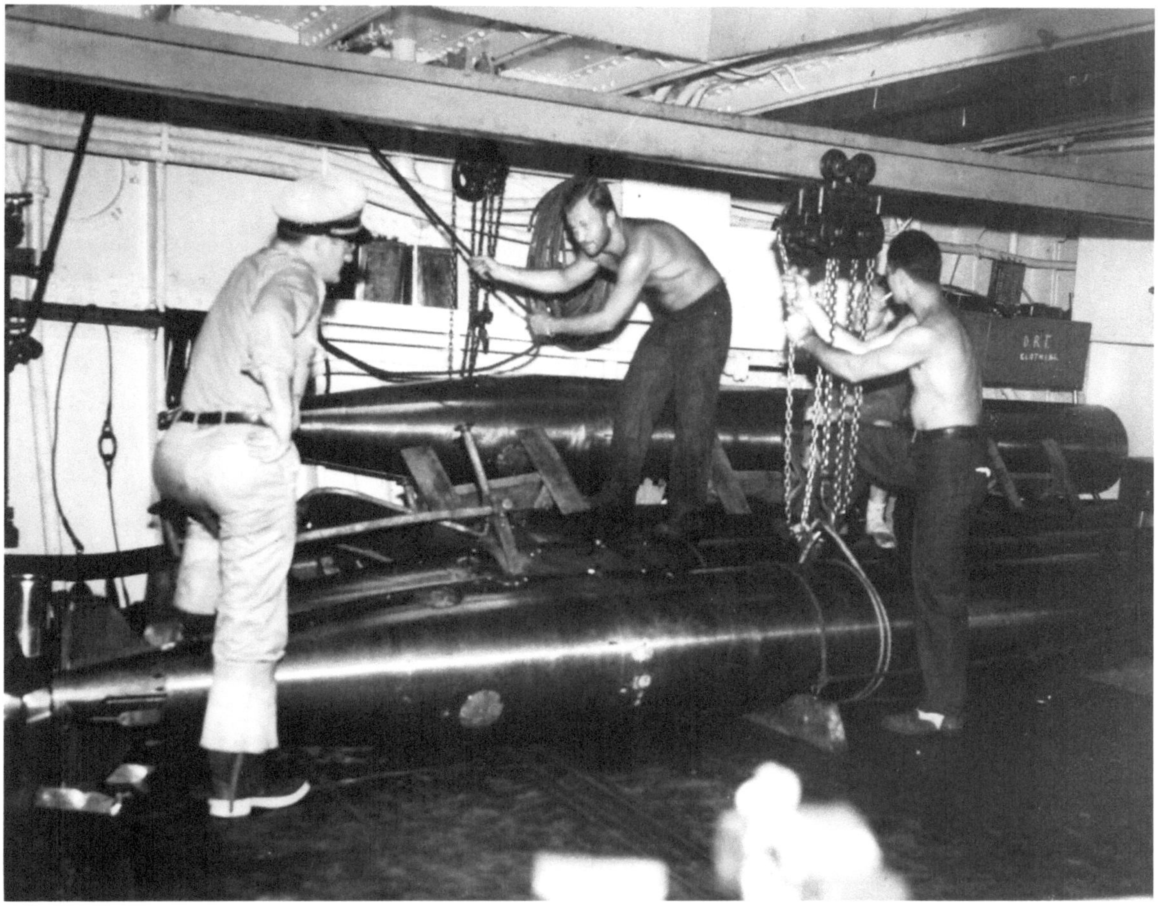

Figure 39. These ordnancemen are busy as they prepare these torpedoes for waiting PT boats. The photo was taken in Leyte, P.I. aboard the tender *Oyster Bay (AGP-6).* Notice the overhead rails in which the torpedoes were attached, then quickly moved to the waiting crane. Attached to the crane, the torpedoes were hoisted up and over the side to the waiting PT boat. (National Archives)

Figure 40. *PT-361* takes on a second torpedo for the aft roll-off rack. With the torpedo high in the air, both men are being very cautious in its handling, as they have rope on both ends of the torpedo to steady it. Notice *PT-332* is also tied to the tender, awaiting service. *PT-361* served with Squadron 27, while *PT-332* was a Squadron 24 boat. (Victor Kodis)

Figure 41. PT tender crew have hooked up this MK-XIII torpedo to the hoist. They are slowly lowering it to the waiting Torpedomen from *PT-361*. Caution had to be used when handling torpedoes in this manner, as one slip could cause the torpedo to fall and bounce off the deck. It will be lowered into the waiting roll-off rack, where it will be secured with cables. (Victor Kodis)

PT BOAT BASES

The development of the PT boat bases was a tiny part of the Navy's vast advance base program. The staggering problem of building bases became simplified by subdividing each into basic components. Faced with the necessity of building hundreds of different types and sizes in all parts of the world, the Navy reduced each base to its simplest form. A functional component is a collection of personnel and material designed to perform one of the specific tasks of an advanced base. It may consist of one man with 100 pounds of equipment, or 1,000 officers and men with 10,000 tons of equipment.

Facilities for messing, housing, defense, communications, and power were not included because they were considered functional components in themselves and provided services to the rest of the base. Seven components were developed as a standard for PT bases. Administration was assigned office equipment, a truck, motion picture projectors, a small boat and office housing. The Operating Repair Base was a semi mobile unit handling major hull repairs, minor engine work, and replacement for one squadron. Major Engine Overhaul provided facilities to four squadrons and assisted other repair units with additional equipment in the machine and carpentry shops. Intended to provide 8 to 12 PT boats with portable, lightweight repair and operating equipment, the Squadron Portable Base Equipment Unit made front-line emergency repairs and worked as a temporary base.

The Ordnance Shop Unit maintained and repaired all armament except torpedoes, which was done by the Torpedo Depot, Small Unit. The Decontamination and Camouflage Unit had no regularly assigned personnel. The Floating Equipment Maintenance Unit (FEMU) which was a conglomeration of people and pontoon barges with a ten-tone crane and two Chrysler Marine pushers, operated as floating dry docks and was formed at Base 21, later moving to Base 17 and Samar. Other examples of the diverse units were Packard Marine Engine Overhaul Unit or PMEOU, which rebuilt complete engines, and the Field Torpedo Unit. In actual practice, standard was rarely seen and varied by addition or omission. Base repair was usually augmented by one or more pontoon piers and dry docks, frequently constructed by whatever was available.

Most PT boat advance bases were considerably smaller than plans had called for. The standard base usually served as a main repair base for several squadrons, while squadrons would usually operate from smaller bases in advanced areas. Main bases were commonly of the standard size and designated by number, 1 to 25, and cared for all squadrons in their geographic area providing repair, supply, staging, and training. Additionally, each squadron had its own Squadron Base Force which moved with the squadron.

SBF services were maintenance, supply, medical and administrative. Advance Bases were shore based and usually supported by a tender. Characterized by a lack of repair facilities and mobility, personnel and equipment came from various sources. Sometimes the advanced base was an individual unit, but more often was a composite of men and material from a Main Base and Squadron Base Forces. PT boat bases, like tenders and the boats themselves, were still under development when the first squadrons were sent to the Pacific. Poor logistics was one of the most serious problems hampering PT boat operations during the first years of the war. The boats were forced to operate from inadequate facilities for a long time.

The base configuration looked like a giant game of leapfrog as advance bases dissolved into more settled facilities as progress was made through the islands. In taking and retaking the islands a unit would move up to support the boats with minimum materials and would be followed when the area was secure with better equipment. Some bases would simply be abandoned while others became very large and remained after the war for a limited time. In Europe, bases were far more adequate because of the already developed nature of the continent. Supply lines existed since the area had been civilized for so long. While other functional components were especially developed for PT's, a typical PT operating base would contain many standard functional components as well. Like the tenders, the bases were not fully developed when the first squadrons were sent to the Pacific, and for many months PT's had to operate with inadequate base facilities. Many of the forward bases were simply makeshift areas of grass shacks, tents, or the boats themselves. In the Pacific where heat and humidity were just as dangerous as bullets, PT boat crews simply slept on deck.

Showers were soapy salt water rinsed off with a bucket of plain salt water. Fresh water was considered sacred, so it was basically used for drinking and cooking purposes. The native population was also used by base force and crews to help with some of the chores, such as rolling gas barrels, unloading supplies, or washing clothes. Once the functional component system was well underway, however, it provided as nearly perfect a solution to base problems as could be imagined, because it was possible merely by adding or subtracting components to form a base suited in size and capabilities for the performance of any required task.

Figure 42. *PT-28* from Squadron 1 is taking on much-needed gasoline. She is at the main fueling dock at Pearl Harbor just prior to the Japanese attack. After the attack *PT-28,* along with two other boats, returned to the Sand Island dock and sent fire and rescue parties ashore. She also participated in the Battle of Midway, and saw service in the Aleutian campaign. She was later wrecked in a storm on January 12, 1943. (National Archives)

Figure 43. This photo shows *PT-42* (a 77-footer) docked at Pearl Harbor. She served with Squadron 1 and was included in the Japanese attack on the base. The boats enjoyed an area that was rich in all the comforts of home. Maintenance on the boats was a constant chore, but was made easy here at Pearl Harbor due to its many facilities. Notice the boat carries two forward MK-VIII torpedoes, a small machine gun on the bow, and several depth charges on her stern. (Author's collection)

Figure 44. This photo takes a look at Pearl City Yacht Club, located in Hawaii. Founded in 1924, it was once visited by wealthy families who spent vacations and summers here. After the attack on Pearl Harbor the US Navy took over this area, which became a small boat base, where PT boat crews and their boats came for additional training. It was here that the Squadrons would await shipment to their designated areas. While at the base, many enjoyed playing golf and formed baseball teams including former major league players. (Author's collection)

Figure 45. PT boat squadrons consumed hundreds of gallons of high-octane fuel daily. The supply of this gasoline for thirsty Packard Marine Engines was vital. In the war zone this fuel was usually shipped in 50-gallon barrels (in the beginning it was pumped by hand). It could take many hours to fill one boat. This photo shows an Army Air Corps gasoline tanker, which was stationed at the base on Taboga. This base was situated just off the Pacific entrance to the Panama Canal, and had many installed facilities, including piers, concrete aprons for hauling boats, torpedo shops, gasoline storage tanks, with living and messing facilities for 125 officers and 700 enlisted men. The PT boats were part of the naval defenses of the Panama Canal, and served as a temporary training base for PT squadrons awaiting shipment to the Pacific areas. (Author's collection)

Figure 46. In April 1943, the PT shakedown detail was established as an adjunct of the Submarine Chaser Training Center at Miami, Florida. Lt. Commander Alan R. Montgomery was the first commander of this new unit. Montgomery worked out an extensive 14 day training program later expanded to 3 weeks. Severe winter weather conditions at the Motor Torpedo Boat Squadrons Training Center led to the decision to conduct all shakedown operations in Miami. Here we see several PT boats docked at this base. (Author's collection)

Figure 47. Officers and crewmembers from the PT boat shakedown base in Miami, Florida, fire a practice torpedo from the newer roll-off racks installed on this boat. Notice the torpedo they are using is the World War One vintage MK-VIII. Usually this type of rack set up was for the newer MK-XIII, a shorter torpedo that was much more reliable and carried more explosive power in its warhead. At the end of its run, the water ballast in the warhead was expelled by compressed air, bringing the torpedo to the surface, where it could be recovered. (Author's collection)

Figure 48. The PT boat base at Tulagi was a simple assortment of tents and bamboo buildings. Most of the base force shops were set up in tents as well as sleeping quarters. Heavy rains would turn these areas into mud infested trenches that made everyday living a nightmare. The base was located in the Solomon Islands, just off the south coast of Florida Island and within sight of Guadalcanal. This base would become a rear staging area when the boats moved to the Russell Islands. (Author's collection)

Figure 49. A small but useful operations shack located at Tulagi. These two Navy men are on station to type orders for boat commanders, dispatch radio communications, and provide other useful office work vital to the base command. (Wallace McNeish)

Figure 50. A native-made grass and bamboo building make up this first aid station at Tulagi. The Aid Station tended to all casualties. Medicines, food, blankets and emergency care were all handled here. Severe cases were evaluated and shipped out to better equipped medal facilities. (PT Boats Inc., Germantown, Tennessee)

Figure 51. A ten-ton crane on this old dock provides the muscle to install one of the powerful Packard Marine Engines. This photograph, taken at Rendova in the Solomon Islands, shows part of what would be known as Base 11. Changing engines was hot and dirty work, but provided the boats with the type of maintenance needed to perform at high efficiency. These powerful engines were scheduled to be changed or overhauled after 600 hours of use. (National Archives)

Figure 52. PT base at Vella La Vella, at Lambu Lambu Cove in the Solomon Islands. The cove was very small in the thick jungles, and the base was nothing more than a few native huts, plus the tents and supplies that were brought. The base improved daily as new tents replaced old ones, a new galley was set up, a shower was made from fuel drums, the jungle was cleared out, and a new head was built over the water. This base would be named Snuffy's Glenn in honor of Lt. Commander Russell H. Smith, USN. (C.J. Willis)

Figure 53. A rare look at *PT-60* (in her gunboat configuration) as she moves slowly out. At the helm is Lt. Leonard (Lenny) Thom, who prior to this photograph served as Executive Officer of *PT-109.* Thom would serve with then Lt.(jg) John F. Kennedy, becoming shipwrecked when a Japanese destroyer hunted, rammed and sank the boat. This would become Thom's second boat after that incident. The photo was taken in Lambu Lambu Cove on Vella La Vella in the Solomon Islands. Notice there was very little room to enter and exit this cove. To the right of *PT-60,* you will see the gasoline dock, which would be destroyed in a gasoline fire on December 15, 1943. (Author's collection)

Figure 54. This is one of only a few known photos to exist of the massive gas fire at Lambu Lambu Cove, on Vella La Vella. On December 14, 1943, at approximately two-thirty in the morning, explosions rocked the night, waking up all those on the Base. *PT-238* and *PT-239* were at the gas dock, with *PT-238* fueling her tanks, and the fuel lines stretched across the deck of *PT-239.* Crewmen from both boats were jumping into the water, as fire started to consume two grass huts used for storing ammunitions. Two members of *PT-238* managed to get the engines started on *PT-239,* but because of the fuel lines across the deck the boat was held in place. An ax was used to chop away the lines, and *PT-238* was pulled away from the dock, while *PT-239* was completely destroyed. Ammunition began exploding, sending tracers in all directions, and fuel drums began shooting into the air. Two men were killed, Chief Gunners' Mate William Conner, and Boatswain Mate Joe R. Mitchell. The two men who cut the lines with an ax, Torpedoman Second Class Richard A. Olsen and Gunners' Mate Second Class Harold W. Foley, received the Navy Marine Corps Medal, for their actions in trying to save the boat. (C.J. Willis)

Figure 55. This photo shows another look at the massive gasoline fire in Lambu Lambu Cove, at Vella La Vella. The smoke could be seen for miles, as the fire burned for hours into the next day. PT boats were subjected to the constant danger of explosions caused by the high-octane fuel used for her three Packard Marine Engines. (Alvin Hansen)

Figure 56. This photo, taken at the advance base located in Morobe, New Guinea, is the scene for crewmembers and base force to lend a helping hand. A sailor works on the aft gun turret spray shield, while a boom has been brought over to replace one of the boats engines. Hot, dirty work was the order of the day under the hot and humid conditions in the Pacific. Notice the hula skirt and laundry hung up to dry, as well as the recognition star that has been painted on the day room cabin roof. These stars were essential so that American pilots would recognize the boats from the air. (National Archives)

Figure 57. This forward base was set up on the banks of the Morobe River in New Guinea. The river was a good choice for the speedy PT boats as it offered deep water right up to the banks and an abundance of overhanging trees, perfect for concealment from the air. The base was established on April 20, 1943 with the arrival of Lt. Cmdr. Barry Atkins with PTs *66, 142,* and *149.* The PT boats ventured out on nightly patrols in an effort to destroy Japanese barge traffic that supplied the Japanese with food, supplies, clothing, and arms. *(*National Archives)

Figure 58. A small Quonset hut sits in the jungles of New Guinea. These buildings played an important part for the base forces, keeping men and equipment out of the elements. Notice in the open door this wing fuel tank for one of the boats. Open windows provided some ventilation, which because of the humid air of the Pacific was almost nonexistent. Still, when the sun was at its highest, these buildings provided much-needed protection. (Gene Kirkland)

Figure 59. Forward PT boat bases were sometimes nothing more than pitched tents and provided many different challenges. This photo was taken at Morobe in New Guinea after heavy rains made travel near impossible. This cargo truck has sunk deep in the mud while trying to haul chopped down trees. (National Archives)

Figure 60. This armor shop was located in Morobe, New Guinea. In the center of the photo, base force personnel are working on 20-mm Oerlikon cannons. This shop stocked hundreds of parts with barrels, ammo canisters, small arms weapons, and more. Notice the Thompson submachine guns to the right of this photo. Weapons were exposed to many elements and needed constant upkeep to insure that they would fire if needed. (National Archives)

Figure 61. Workers fill and stack sand bags on Base 9. This base was located on Cape Torokina, Bougainville. Notice the aircraft radar warning system they have set up, which is one of the PT boats' SO radar masts. The small truck to the right of the photo contains the generator that runs the power to the radar dome. (National Archives)

Figure 62. This is the metalsmith and shipfitters' shop, located at Torokina Base, Bougainville. This simple tent structure with dirt floor is providing the sailors with some protection from the elements. Left to right: W.B. Bradley, Ship Fitter Second Class; O.G. White, Motor Machinist Mate First Class, and R.C. Steen, Construction Mechanic Third Class. Bradley is using a torch to cut metal, while the others are busy performing other tasks. Notice the worktable space and other equipment used for repairs. (National Archives)

Figure 63. In this photograph taken at Torokina Base, Bougainville, H.M. Garothorp, Baker Second Class is kept busy preparing and making fresh hot bread. These ovens were made from empty oil drums, a simple but effective means for cooking. Heat and humidity, not to mention the constant appearance of bugs and small rodents, made working here a tough chore. Cooks worked hard to prepare hearty meals for the base forces and PT crews alike. (National Archives)

Figure 64. Life on many of the bases (especially in the forward areas) was made difficult by weather conditions. Heat, humidity, driving rain, bugs, and of course the enemy, were some of the major discomforts to the base forces and PT boat crews. If one was lucky, he could get along at the base with a jeep or truck. Although having a vehicle was a luxury, these sailors might not think so, as they are broken down on one of the major roads on the Island. (Alvin Hansen)

Figure 65. A simple but effective repair tent, set up on Base 7, Green Island in the Solomons. Inside, Holly carburetors are being cleaned, fixed, and adjusted. When completed, they will be installed on Packard Marine Engines ready for testing. Many times a shortage of parts and supplies would hamper the boats ability to run at top speed. Parts from other boats would be used to keep as many boats on patrol as possible. (PT Boats Inc., Germantown Tennessee)

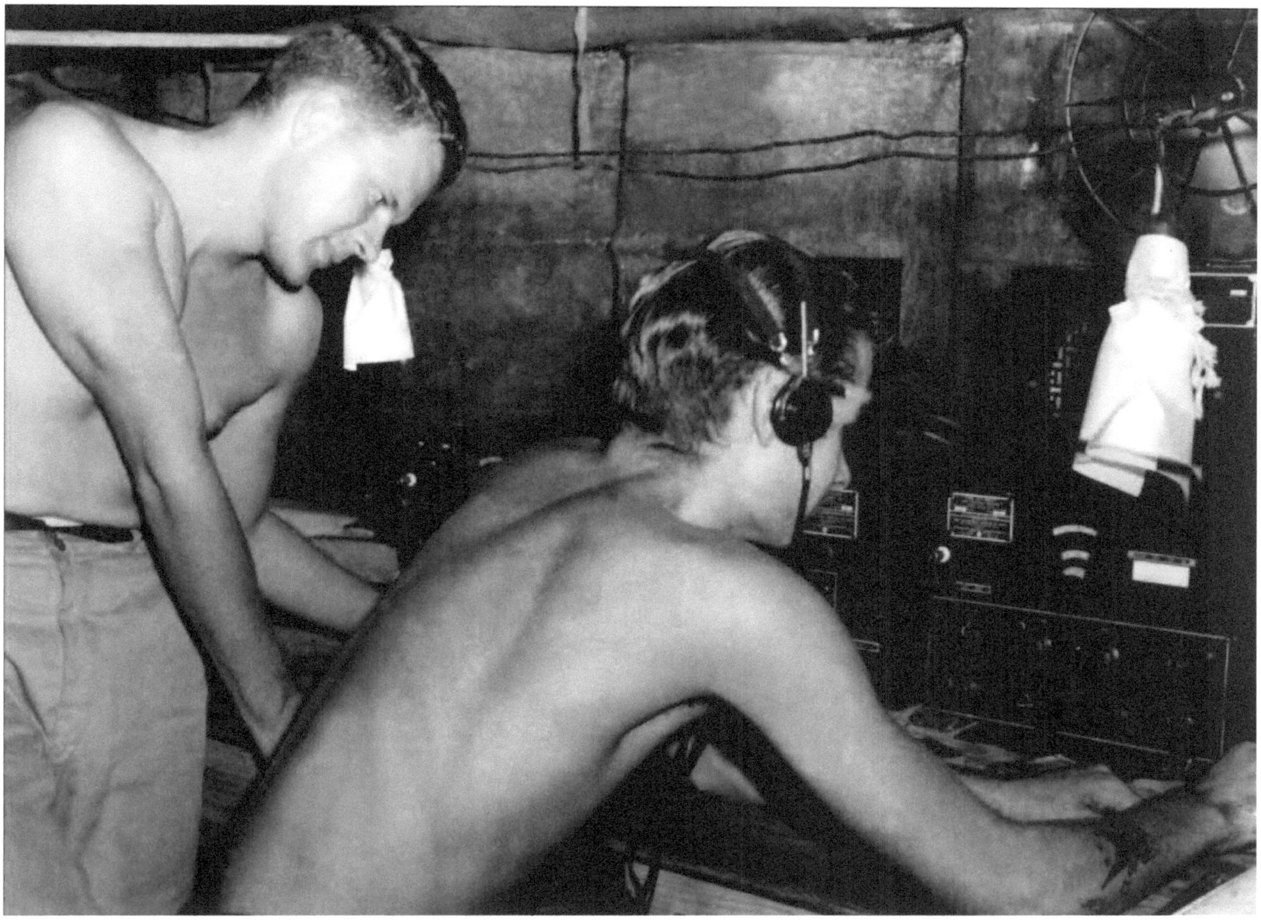

Figure 66. The radio communications center on Green Island seems to be busy, as these men are listening in on Japanese movements around the area. Reports will filter in, and will be given to the squadron commander, who in turn will get them to the boat captains. A briefing will follow, and selected boats will be given their assignments for the nights patrol. (C.J. Willis)

Figure 67. This photo shows the jungles somewhat cleared out. The buildings that were built here are the carpenter shop (left) and the engine overhaul shop. The truck was used to haul supplies and other parts from the dock to these small buildings. This was at Base 7 (Green Island) in the Solomon Islands. Base force personnel were kept busy trying to keep the boats operational, and at one point the tender *USS Varuna* stood by to assist with repairs, and other needs of the Base, while it was under construction. (C.J. Willis)

Figure 68. The medical staff on Green Island stands next to their medical tent. Supplies were brought in by supply ship, although some supplies were originally brought in when the base was set up. All medical emergencies were treated here, with the more serious cases being moved to more advanced medical facilities. (C.J. Willis)

Figure 69. This area was cleared out on Base 7, and was used for entertainment and church services. Logs were constructed to act as seating, and in the beginning movies were viewed once a week, then every night as more movies were shipped in. Here, Sunday church services are taking place. (C.J. Willis)

Figure 70. Base 7 on Green Island was the setting for this Quonset hut. Nicknamed the "Greasy Spoon", this was the ship's store where men from the base and the boats could purchase candy and tobacco. Prior to these huts being built by the Seabees, the base was nothing more than leaking tents with boats moored to bushes and floats. Docks would be built later on for the boats. The identity of the officer standing in front of the hut is unknown. (C.J. Willis)

Figure 71. This small duty boat was kept busy on Green Island, as side trips to other small Islands provided much-needed information from area natives. The boat was also used for fishing, which was very good in the area with sharks, barracuda, and other tropical fish that provided many hours of enjoyment. The boat would also be used to run out to pick up the mail, something everyone on the base looked forward to. (C.J. Willis)

Figure 72. This officer, with base force help, is checking this torpedo by using a tripod stabilized valve-lapping device. A large quick-opening air valve at the top rear of the tube allowed the entire air reservoir to dump into the torpedo tube at the stern end of the torpedo when it was being fired. Making sure that there was a good seal on this quick-opening air valve would be critical to the proper operation of the torpedo. By the looks of this photo, they are trying to repair the sealing surface of this valve by lapping the valve seat. This job required patience and lots of elbow grease. (C.J. Willis)

Figure 73. This sailor is busy adjusting the clamp at the aft end of this torpedo tube. This end of the tube holds the end bell on, and he has just finished working on the propellers of the torpedo inside the tube. Later he will remove the torpedo lock to inspect that it is working properly. Torpedoes that do not fire in combat are useless, and maintenance was performed on all working parts of the torpedo. (C.J. Willis)

Figure 74. On December 28, 1943, *PT-242* was off the coast of Bougainville when they ran into some Japanese barges. After closing in to roughly thirty yards, they commenced firing, and received return fire from the Japanese. During the ensuing gun battle the barge was destroyed—and then it was noticed that the PT's forward port torpedo was smoldering in the tube. The torpedoman (John Grace) fired the air charge, trying to release the torpedo, but it was wedged in the tube. It was later learned that a stray bullet had hit the warhead and split the metal casing. The torpedo and torpedo tube were damaged beyond repair. Here are a base force torpedoman (right) and crewmember look at what is left of the tube. (C.J. Willis)

Figure 75. The Bob Hope Troop arrives at the Airport. Left to right: comedian Bob Hope, comedian Jerry Colonna, dancer Patty Thomas, guitar player Tony Romano, and singer Frances Langford. PT boat crews and base force were entertained on July 1, 1944, when Bob Hope and his troop came to Green Island. The troop has just landed at the airport and shortly will be escorted over to the base. (C.J. Willis)

Figure 76. *PT-237* had the pleasure of taking the Bob Hope Troop over to Green Island. They would pick up the group at the fuel docks at the airport, and take them across the bay. Executive officer of the boat, Ensign Small (left), is on top of the world as he has the pleasure of leggy dancer Patty Thomas (right) on the boat. (C.J. Willis)

Figure 77. The enemy doesn't have a chance as Bob Hope (left), Patty Thomas, Tony Romano, and Jerry Colonna man the twin .50 caliber machine gun set up on *PT-237*. During the show on Green Island, it started to rain, and singer Frances Langford asked the men in the crowd, "What do you do when it starts to rain?" They all answered in unison, 'We get wet." This broke up the crowd, and the show continued with all those in attendance getting soaked. Entertainment at the front lines was so important to the men, which brought them a piece of home to their otherwise rugged daily routine. (C.J. Willis)

Figure 78. A look at the inside of the torpedo shop, located in Mios Woendi, which would be known as Base 21. These torpedoes are being worked on to insure smooth operations when used in actual combat by the boats. Notice the torpedo-carrying carts on wheels, which in this case are holding the war heads of the torpedoes. These are MK-XIII torpedoes, which were used in conjunction with the newer roll-off racks, installed on most boats later in the war. (PT Boats Inc., Germantown, Tennessee)

Figure 79. A look at one of the metal lathes used at Base 21. Overseas bases needed trained machinists to operate these delicate pieces of equipment. Many of these operators went to school prior to joining the Navy, or already had jobs in civilian life. They would take that much-needed experience and put it to good use making parts and equipment for the boats. (Randall J. McConnell)

Figure 80. Base 21 shows an inside look at the metal shop. PT boats in the war zone were in constant use and required an extensive maintenance program. Although primarily a wooden boat, metal parts such as shafts, struts, props, and gears were in need of repair. The metal shop provided machines and trained men that could make repairs to keep the boats in running order. (PT Boats Inc., Germantown, Tennessee)

Figure 81. Base 21 also had a welders shop as seen here. These welders are using oxy-acetylene equipment, which could be used to cut metal or steel pieces. Spot welding was also done in the welders shop, with an electric arc welder. Welders were in high demand during World War Two. Welders needed good eyesight with the ability to read blueprints. Work at the welders' shop not only included items for the boats, but all other welding needs for the base. (PT Boats Inc., Germantown, Tennessee)

Figure 82. PT boats carried many different weapons and it was the job of the base force and gunner's mates to make sure all weapons and parts were always in working order. This sailor is taking inventory of the many hundreds of parts that make up this armor shop. Notice the .50-caliber machine gun barrels as well as the .50-caliber machine guns, and .30-caliber machine guns and parts that line the wooden bins. This photo was taken at Base 21 in Mios Woendi, New Guinea. The base was used for supply, staging, repair, and training purposes. (PT Boats Inc., Germantown, Tennessee)

Figure 83. Here cooks prepare a hearty meal at Base 21. Fighting men needed a good diet, and from the looks of the meat in this galley, no one would go hungry. Seabees constructed this Quonset hut, complete with running water for the sink, cooking facilities', and refrigerated food lockers with generators. It should be noted that not all PT boat bases had facilities such as these. Many were simply tents, with the most primitive of cooking facilities. Supplies were not always available, and Spam™ was the order of the day. Many boats would race out to tenders or other ships, to beg food and supplies for the men. (PT Boats Inc., Germantown, Tennessee)

Figure 84. This torpedo shop (complete with concrete floor and water drainage) shows a comfortable working space. The sailors in the middle are working on the combustion flasks. Notice the stack of torpedoes to the left of the sailors contains just the propulsion motor and the oxygen and fuel flasks. No warheads are visible, which means they were stored in another area. These are known as MK-VIII torpedoes; they were of World War One vintage. The sailor to the left of the photo is brushing on peralkytone, a Cosmoline™ type mixture (only thicker) to protect the torpedo from rust. (Author's collection)

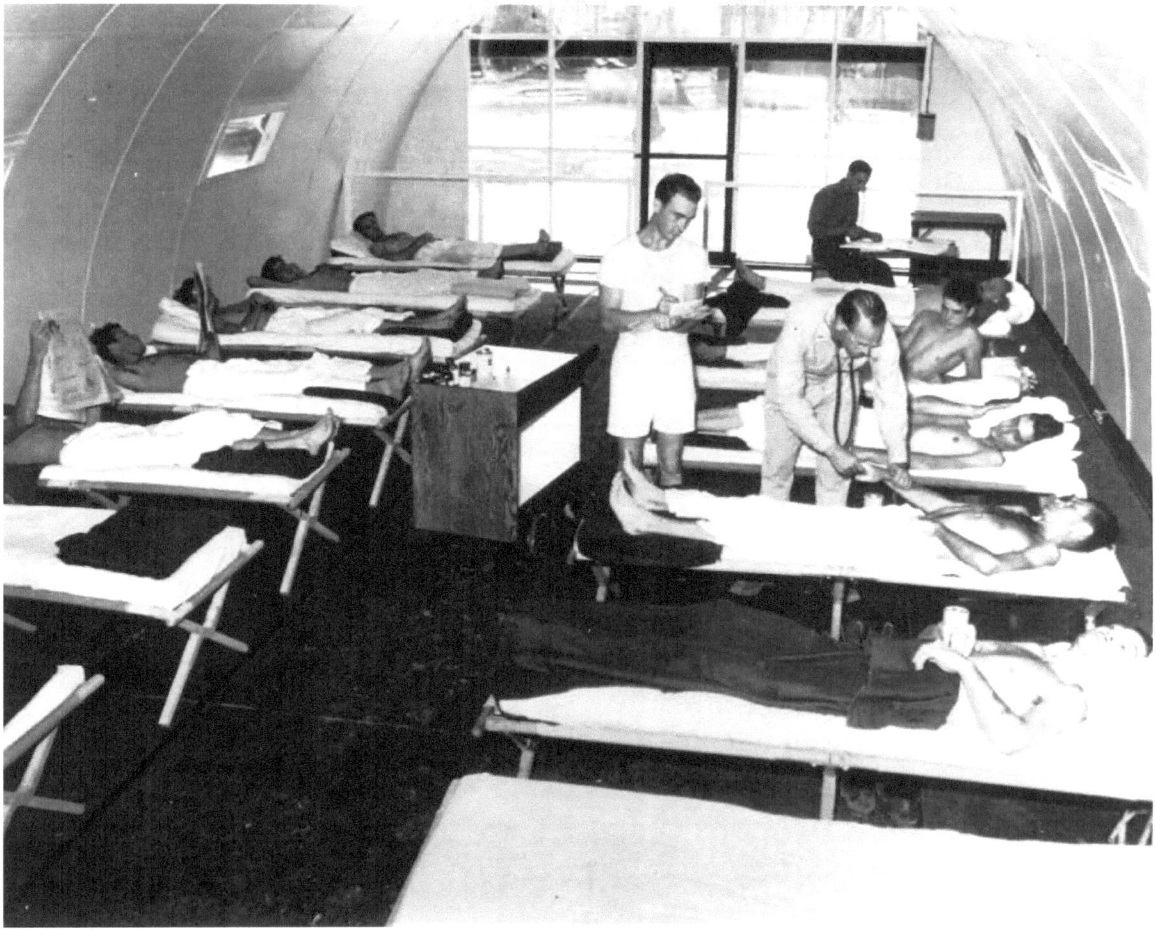

Figure 85. Here we see another medical facility in the War Zone. These young Sailors are being tended to by a Medical Officer and his assistant. Many of the illnesses were caused by mosquito's carrying Malaria, which caused high fever, and chills. Tropical ulcers and cholera were some of the other illnesses that cropped up. Most were treated by pills, powders, or injections, with many Sailors returning to active duty. (PT Boats Inc., Germantown, Tennessee)

Figure 86. Base 21 wood machine shop workers are hard at work. Much of the wooden parts from the boats would be repaired or made here. Planks for Hull work, patches, cabin trunk structural work, and decking would be provided. Jig saws, wooden lathes, band saws, and hand tools would be kept in this area. This shop was kept busy throughout the course of the day also providing fabrication work for the boats. (PT Boats Inc., Germantown, Tennessee)

Figure 87. This photo shows a happy bunch of sailors who have jumped onboard this vehicle. Base force personnel needed trucks and other vehicles for hauling supplies from one place to another. At times, heavy rains on the Islands could make driving next to impossible. From the looks of this small truck, it would seem to get plenty of work. (Robert Douglas)

Figure 88. PT boats required a regular program of maintenance to keep them on patrol and in shape. Engines were tuned, guns were oiled and fixed, and at times deckhouses and other fixtures would be scrapped and painted. Here we see two men from Advance Base 4 at Morotai getting ready to do some painting. One holds the brush while the other does the mixing. Paint was secured by the supply hut, and generally was whatever color was in the can (as close to the boats color as possible). (Author's collection)

Figure 89. A look at the inside of this building shows the engine test bed for one of the 4M-2500 Packard Marine Engines. This was Base 4, located at Kana Kopa on Milne Bay. It would become a major engine overhaul base for PT boat operations until moving to Dreger Harbor in 1944. Tired engines would be brought here for repair and overhaul to include cylinder banks, pistons, crankshafts, and connecting rods. Before overhauled engines were installed on the boats for everyday operations, they were carefully inspected and run according to procedure. Fresh water, fuel, and oil tanks are visible here. (PT Boats Inc., Germantown, Tennessee)

Figure 90. This Pacific based torpedo dump shows the many torpedoes stacked and ready for installation. A tracked tug carefully extends its boom to place this MK-XIII on the waiting torpedo cart. Torpedo bodies (propulsion motor and oxygen and fuel tanks) were stacked outdoors, with the warheads stored indoors. Warheads would be secured to the bodies of the torpedoes, and then brought over to the boats for installation in tubes, or roll-off racks. (Author's collection)

Figure 91. Some games were played at the bases when time permitted, and were usually when an area was secured. Here these dark tanned men from Green Island get together for a good game of basketball. Notice the usual dress of those who served in the Pacific area was shorts, combat boots, and no shirt. (C.J. Willis)

Figure 92. Although life at any base could sometimes be considered routine, it was important that some type of activity bring a piece of home to the boys. These Squadron 21 base force members have not only taken up singing to pass the time, they have managed to scrounge up a working piano. From the looks on their faces, they are having a good time. (PT Boats Inc., Germantown, Tennessee)

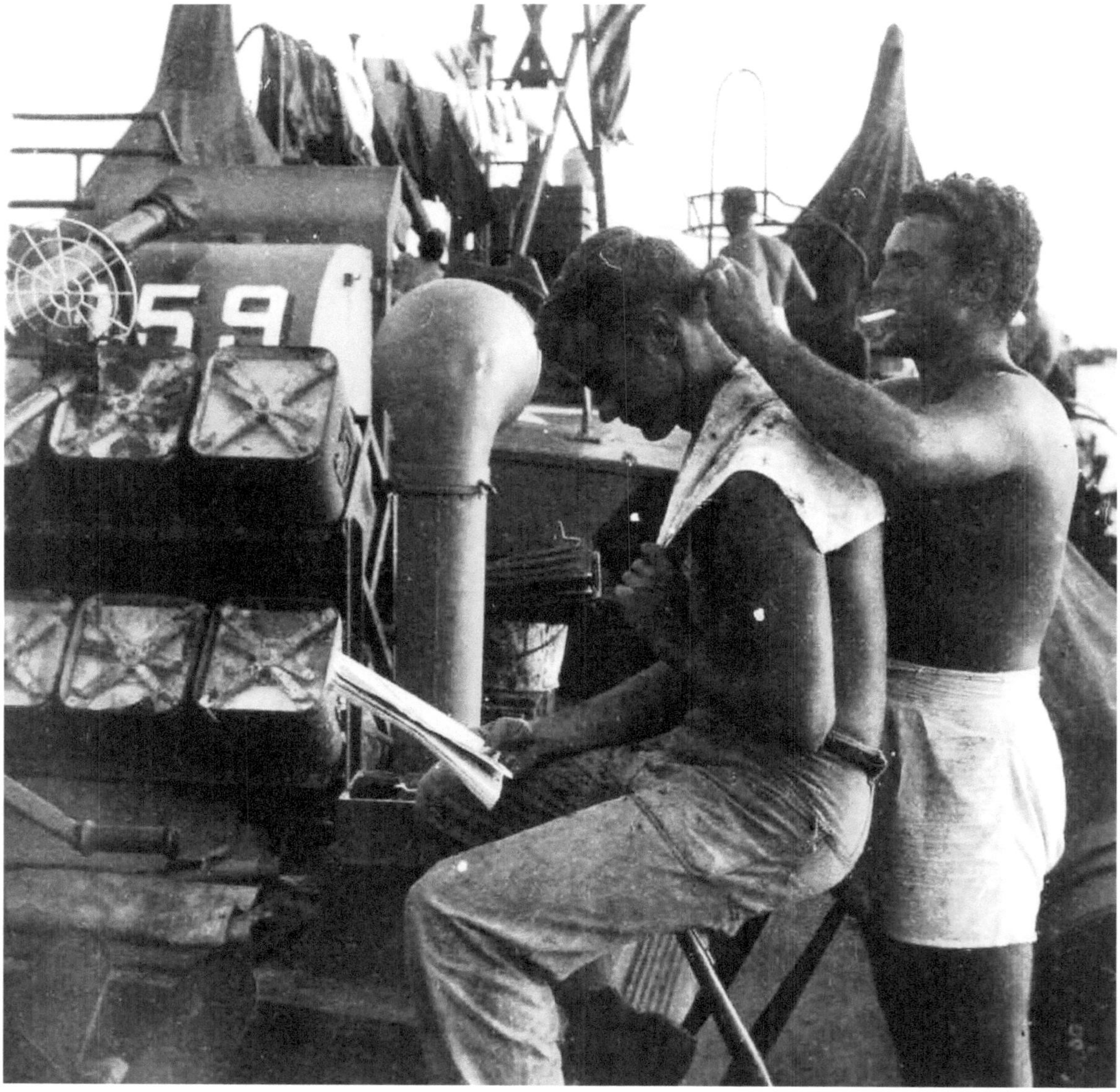

Figure 93. Being in a forward base with limited supplies and washing facilities, Base force personnel and crews alike did what they could to maintain their personal hygiene. It was not uncommon for someone to grab a pair of scissors (as shown here) and cut hair. Hot showers were out of the question on most bases, although supplies like soap, razors, foot powder, and shampoo could be purchased on the larger bases. (Alvin Hansen)

Figure 94. Bathing for the common PT Boater was a tough chore (on most bases), as fresh water was used very sparingly. It was commonly used for drinking purposes, and not wasted on showers or baths. This was where the unpopular salt water bath came to the front. Salt water soap was used to lather up, than the bather simply rinsed off with salt water, using a bucket or a quick dip in the ocean. As some Pacific bases became more established, Seabees installed fresh water condensers to help with the water shortages. (United Press Photo)

Figure 95. Maintenance crews will be kept busy, as they will have to replace the barrel of this 40-mm Bofors cannon. It is unclear as to how this might have taken place. Possible causes could have been a shell that was hung up in the barrel, or an incoming round from enemy fire. Luckily, no one was injured on the boat. (Alvin Hansen)

Figure 96. *PT-297* sits swamped on the beach after being attacked by Japanese planes. The boat was attached to Squadron 16, based at Mangarin Bay, Mindoro. Operating with PT-224, the boats were departing for the night's patrol, when they were jumped by two planes. Three bombs were dropped from the Japanese attacking planes and the concussion caused PT-297 to settle at the stern before resting on the beach. Base Force crews would strip her of all necessary equipment to try and keep her afloat. (National Archives)

Figure 97. PT-146, nicknamed "Lucky Lady", has managed to intercept the escort carrier *Sitkoh Bay (CVE-86).* Serving with Squadron 12, the little PT boat looks even smaller as she is dwarfed by the large ship. Her skipper has tied up to the carrier in attempt to beg food and goodies for the crew. Many times sailors from these large capital ships would hug the rails to catch a glimpse of these little midget sluggers. In some instances food was exchanged for a quick ride around the harbor, often with those taking the ride becoming sea-sick. (Robert James Douglas)

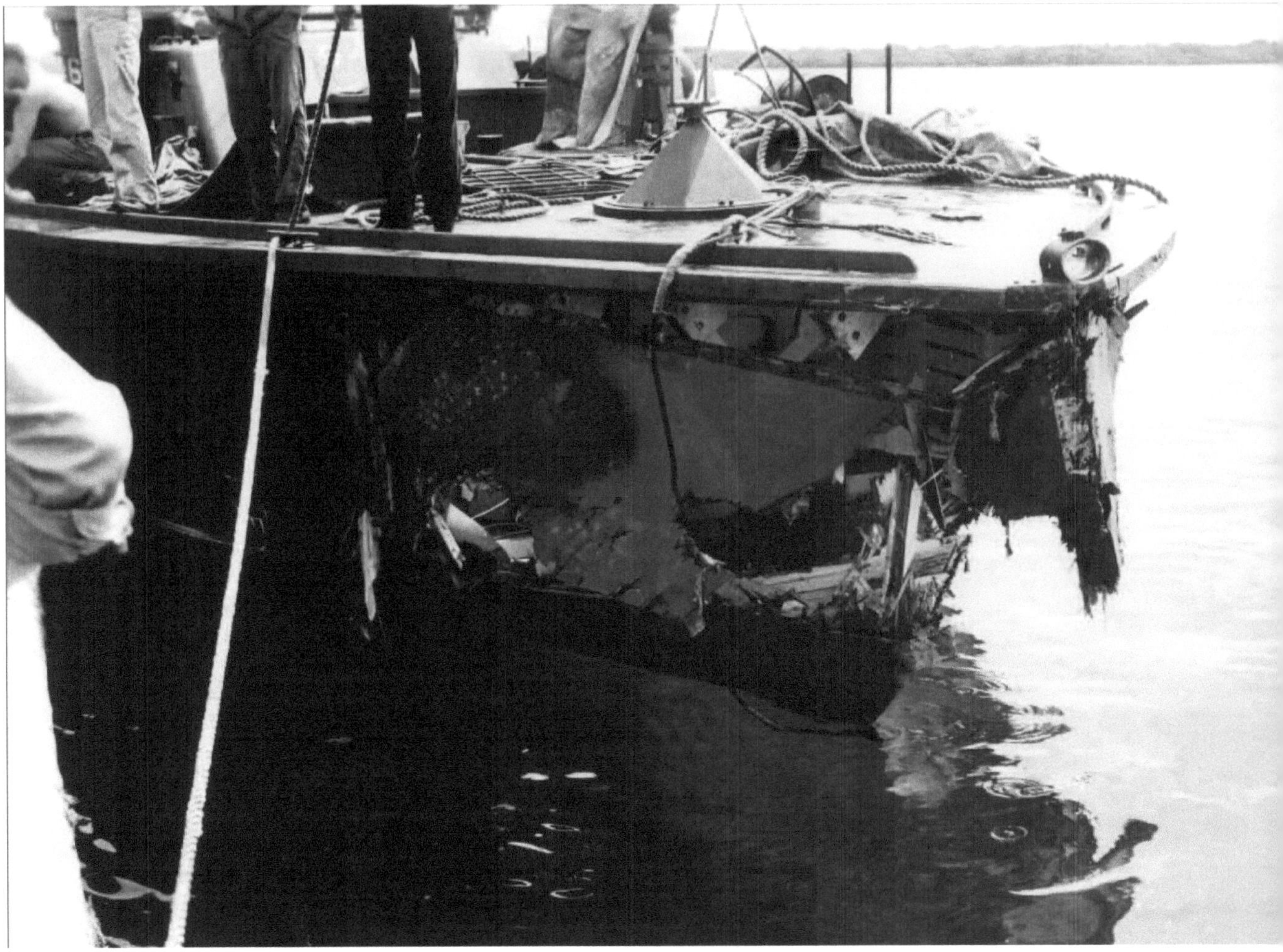

Figure 98. Although the boats were built to withstand the rough twisting and turning of the ocean, they were still made of wooden construction. As seen here, the carpenters of the base force will have their hands full fixing the bow of this Elco boat. (Author's collection)

Figure 99. These Squadron 15 boats are lashed together at the dock at Meddalena, Sardinia. This was an advanced shore operating base. The already established buildings in the area made things easier for the crew. One of the blasted buildings was repaired and became the repair shop for the boats. This base was also shared with the British Coastal Forces. These boats still carry the outdated 20-mm cannons on the stern. (National Archives)

Figure 100. *PT-216,* a Squadron 15 boat, sits on the dock at Base 12. This was the main PT boat base in Bizerte, Tunisia. This would be the main base for Packard Marine Engine overhaul, right until the end of the Mediterranean campaign, and was located about 100 miles from Bone. Notice this Higgins boat is being fueled with the use of a portable pump, cutting down on the time it took to empty the fifty gallon drums to the left. (National Archives)

Figure 101. Three brand new 4M-2500 Packard Marine Engines sit on the dock at Base 12, Bizerte, Tunisia, awaiting installation on one of these Squadron fifteen boats. Base force personnel to the right of the photo take a much-needed break, before installation begins. Installation of engines would easily consume an entire day or more. Notice the zebra striping on *PT-208* (in the background), which carries the earlier 20-mm cannon on the stern, as well as the early style torpedo tubes. What appears to be an early style rocket launcher sits on the bow. (Author's collection)

Figure 102. The inner harbor at Bastia, Corsica, shows the many boats that were based here These are Squadron 15 boats, which would encounter German surface targets from the enemy held island of Elba. Moving to this base would bring the entire Gulf of Genoa within easy patrolling distance, but finding shelter ashore would prove difficult. In the beginning, all but the most minor repairs had to be done at Bizerte, but worst of all torpedoes could not be maintained properly until base force equipment arrived from Palermo. (PT Boats Inc., Germantown, Tennessee)

Figure 103. Base force personnel were certainly kept busy as this photo will testify. PT-307 has considerable damage to her chart house area. While patrolling between Leghorn and La Spezia, PT-307, 304, and 306 encountered an enemy destroyer and a corvette, and gave chase. In the ensuing battle, PT-307 was raked with 20mm and 40mm gunfire. Three men were killed and five wounded, as the boat made its way back to Bastia, some 90 miles away. (National Archives)

Figure 104. A look at the considerable damage to the bow of *PT-303.* Orders were received to load six MK-VIII torpedoes on *PT-302,* 303, and *305*. The mission was to attack the sea wall at Genoa, Italy. These older torpedoes were used simply to get rid of them, and could only be carried in the boats forward roll-off racks. Orders were for *PT-303* and *305* to fire the torpedoes and turn outward away from the sea wall. In the confusion, both boats turned inward with *PT-303* slamming into *PT-305,* and crushing in a good portion of her stern. With water tight doors closed, and speed reduced considerably, both boats made it back to their base in Gulf Juan. With an excellent base force, both boats were back in operation in just over one week. (Arthur Frongello)

Figure 105. A look at the damage caused to the stern section of *PT-305.* Notice the bracing used to add strength to the stern when the boats added the 40-mm Bofors cannon. (Arthur Frongello)

Figure 106. A wonderful look at Squadron 29 (front) and Squadron 15 (nested against sea wall) at their base in Calvi, Corsica. This base was on the Western side of Corsica, where the boats would be closer to the French coast and the Italian coast west of Genoa. At the time this photo was taken, the boats had already had their heavy torpedo tubes changed over to the light weight roll-off racks. What is interesting is *PT-553* (third from left) is carrying MK-VIII torpedoes in her forward racks. Notice the large sea wall that helped to provide protection from both the enemy and the sea. (Author's collection)

Figure 107. Another beautiful look at Squadron 29 and Squadron 15 boats (docked) at their base in Bastia. This base was shared with British Coastal Forces. Notice the dock area now has several buildings, including a Quonset hut, several supply trucks, torpedo crane, and other equipment Base Force crews would need for their operations. (PT Boats Inc., Germantown, Tennessee)

Figure 108. *PT-306*, nicknamed "The Fascinatin' Bitch", is tied up alongside this unknown ship. She served with Squadron 22 in the Mediterranean campaign. While part of the crew poses for the camera, others are next to the charthouse playing cards. Notice the crew's laundry hanging on a rope. It is possible the boat has tied up alongside to try and obtain fresh vegetables, food, or even cold beer. Food and supply shortages were not as common in the Mediterranean as they were in the Pacific. (Author's collection)

Figure 109. Even when docked at the base, many crews simply choose to eat their meals on the boats. This was especially true if the boat's cook had a flair for making good meals. This crew has managed to secure a table set up (complete with makeshift chairs) so as to have their chow topside. By the looks of this crew, they seem to be enjoying what has been set before them. Many bases offered a decent variety of food for both the PT boat crews and the Base Forces, while a lack of supplies at others caused crews to beg and borrow where they could. (Bill Pleasant)

Figure 110. Sailors crowd in front of this occupied building nicknamed "The Mosquito Mansion." They are part of Squadron 22 based in Golf Juan, France. Those squadrons in the Mediterranean for the most part lived in better conditions than their counter-parts in the South Pacific. This was simply because there were existing buildings for them to use as bases. (PT Boats Inc., Germantown, Tennessee)

Figure 111. *PT-302* takes on an interesting torpedo. Crewmember Arthur Frongello (hand on torpedo) helps with this MK-VIII. Because of a shortage of torpedoes, this older model was given *to PT-302* by an American submarine captain. Much longer than the MK-XIII, the boat would carry two of them in the forward roll-off racks. They would later fire both torpedoes at the Sea wall in Genoa, Italy. (Arthur Frongello)

Figure 112. *PT-302* from Squadron 22 has been put on the marine railway of this one privately owned yacht club. This photo was taken in Gulf Juan, France, where the boats were pulled up on the railway to have their hulls scraped and painted. Repairs to the hulls, shafts and screws were also be completed. Although much of the hull scrapping and painting work was done by the crews, base force personnel were called in to complete more serious repairs that would pop up, requiring replacement. (Arthur Frongello)

Figure 113. Base force cooks have done a wonderful job creating this anniversary cake. It was the first anniversary for Squadron 15, which was the first PT squadron sent to the Mediterranean Theater, where it operated as a unit of the British Coastal Forces. No doubt this cake did not last long once the crews got wind of it. (PT Boats Inc., Germantown, Tennessee)

Figure 114. A pretty girl with her twirling Baton takes time to chat with members of Ron 22. The boat she is in front of is PT-308, nicknamed La Dee Da. She is part of the entertainment for the troops sent from the USO. For most of these men seeing a pretty girl from home was a good enough reason on why they were fighting this War. (Walter Pirog)

Figure 115. The PT boat Base in Casco Cove, Attu. Squadron 13 participated in the Aleutian campaign from March 1943 to May 1944. Squadron 16 also became part of this cold weather base beginning in August 1943. Patrols from this area proved to be negative, as the Japanese never again sent ships to this area. The boats also helped ships through the fog from Massacre Bay to Shemya Island, where an air strip was being constructed, and transported high ranking officers to the northern side of Attu. The small building to the right of the dry dock was used for maintenance. Extremely cold weather and high seas hampered the little boats, making life here difficult. (PT Boats Inc., Germantown, Tennessee)

Figure 116. A look at the inside of the torpedo shop on Base 5. This small shop was located in Adak, Alaska. Notice the very efficient bench set up where valve testing was completed. Through the door was the gyro room. A small coal burning stove against the wall was the only means of keeping warm in the cold climate of Alaska. (PT Boats Inc., Germantown, Tennessee)

Figure 117. No area was as cold as the forward bases in Alaska. Cold temperatures combined with high winds and rough seas made life here miserable. Quonset huts with small pot belly stoves were protection against what became known as "williwaws." These were squalls that would sweep down off the mountains with terrifying speed, building up to gale winds of 100 knots. This photo was taken at Base 13 in Attu, and no doubt depicts all the comforts of home that were available to those living here. (PT Boats Inc., Germantown, Tennessee)

Figure 118. *PT-28* from Squadron one sits on a marine railway in Dutch Harbor Alaska. She is undergoing hull repair after going aground. Notice the early aircraft radar that has been installed on the bridge area of the boat as an aid to navigation in the Aleutian fog. On the night of January 12, 1943, PT - 28 would attempt to pull PT-22 off a reef. The boat went hard aground and after several attempts by YMS-125 would free her. The tow line broke and PT-28 washed up on the rocks on the northwest shore of the harbor. All further attempts at salvage failed and PT-28 was abandoned and later broke up on the rocks. (National Archives)

Figure 119. *PT-508* sits at the dock in Portland, England. The ten-ton crane on the dock is maneuvering a new engine into place. This main base was occupied by both British Coastal Forces and American Forces. Many of the PT boats came here for repair, staging, and supplies. The PT boats were not needed in the English Channel until the spring of 1944, when they were asked to join the screening forces in the invasion of Normandy. Notice the many different types of small craft in the harbor. (PT Boats Inc., Germantown, Tennessee)

Figure 120. These men from the main PT boat base in Portland, England stand in front of the GSK Quonset hut. This was the General Store Keeping Facility where parts, tools, and other vital supplies were kept. Among these Quonset huts were administrative offices, post office, and Squadron offices. (Donald Millman)

Figure 121. The quartering and messing area for the PT boat Squadrons in Portland. This would become known as Tent City, and was located roughly one mile from the repair center, midway up the side of a hill overlooking the Channel. There were ninety-nine tents, housing approximately 800 officers and men. The site also included a sick bay tent, two wash and shower buildings, two storage tents, a laundry tent, two food storage tents, a large combination mess hall and movie hall, other areas for officers, and a concrete-decked galley building. The area was adequate in the summer months, but became a rugged area to live in during the fall as high winds and driving rain would blow down tents. The tents were heated with coal stoves, which proved to be a fire hazard. (Donald Millman)

Figure 122. These hungry sailors from this base in Portland, England have formed quite a long chow line. Weather was a constant problem in the fall, as driving rain and high winds caused havoc with these tents, blowing them down. It took a great deal of supplies to keep these men well fed. (PT Boats Inc., Germantown, Tennessee)

Figure 123. Notice the large wooden repair shop at Portland. Base force crews worked around the clock in an effort to keep the squadrons up and running. Everything from hull repair to upper deck structure replacement was conducted here. The Base force crews shown here were very skilled and produced some wonderful carpenters, who could put a boat with extensive damage back in the water in only a few days. (Don Millman)

Figure 124. Transportation was vital on any base. Here two Dodge WC-51 trucks have pulled into park at the Portland base. Notice the first truck has a couple of PT crewmen in the back. These vehicles were vital for transportation of parts and equipment, especially during the rainy season, which made life miserable at the base. Base force mechanics saw to it that the trucks were well taken care of, as a broken truck meant someone might be hauling things by hand. (Don Millman)

Figure 125. *PT-505* (Squadron 34) plies the waters around Normandy. On the night of June 7, 1944, Lt. William C. Godfrey gave chase to what appeared to be a submarine periscope cutting the water near St. Marcouf Island. Godfrey was about to give orders to drop depth charges when the boat hit a mine. The explosion lifted the stern out of the water, injured two men, tore loose one depth charge, snapped the warheads off the torpedoes, threw the engine beds awry, and caused damage to just about every part of the boat. She went down by the stern until the 40-mm. cannon was awash. Godfrey jettisoned his torpedoes and his other depth charge, and transferred his forward guns, radar, and radio equipment to *PT-507*, which towed the *505* to anchorage in the lee of St. Marcouf Island. (Donald Millman)

Figure 126. This photo shows *PT-505* after being towed to Portland by *PT-500*. After hitting the mine, two LCMs towed the stricken boat onto the invasion beach at high tide. At low tide, the crew put emergency patches on the hull. Here officers, crewmen, and base force personnel at Portland survey the damage. Notice the boat has been stripped of her 40-mm Bofors cannon on the stern, while workers are busy in the engine room dismantling the engines. (Donald Millman)

Figure 127. *PT-505* minus her engines after the boat hit the mine. Engineers dismantled the three Packard engines and removed them from the boat. Base force personnel made the necessary hull repairs, including replacing the double planked mahogany under the engine room, as well as other parts of the hull. (Donald Millman)

Figure 128. *PT-459* from Squadron 30 sits in the repair shed in Portland, England. These sheds were instrumental in supplying a somewhat dry environment for the Base Force crews to carry out their repairs. In the lower left of the photo you can see the workers busy cutting and fabricating whatever parts were needed. Repairs other than wooden hull work was also conducted here. (Donald Millman)

Figure 129. The elements such as salt water spray, humidity, cold, and weather all played a factor on the wear and tear of the boats deployed in European waters. Topside fixtures such as day room cabin, chart house, spray shields, ventilators, and more needed repair and painting. Here this sailor is busy painting the ammo lockers of this boat. (Donald Millman)

Figure 130. Base force crews at Portland are kept very busy as they strip a good portion of this PT boat's bow section. They have stripped off the damaged Mahogany planking and will complete a full front restoration. You can see into the crew's quarters of the boat looking aft. Work at this base was done under cover of the boat repair sheds. Notice the wooden staging that has been erected around the boat. (Don Millman)

Figure 131. *PT-517,* a Squadron 34 boat, has been put into her cradle. The boat has been hauled onto the concrete apron at the PT boat base in Portland. Some of the crew is busy squaring things away topside while base force workers are starting to survey the planking on the bow. The boat will undergo a complete examination of the hull and any damage will be repaired. The hull will then be scraped and painted, and the struts, shafts, and props checked for damage. If need be the Packard Marine Engines will be changed (notice the engine shipping crates to the left of the boat), which would leave the boat in drydock for at least a week. (Don Millman)

Figure 132. Tension on the base sometimes reached critical levels for both base forces and PT boat crews. Here the boys take up a game of volleyball to pass the time in Portland. These activities were also vital for physical exercise. It was not uncommon for side bets to be made for some of these games between boat crews. (Don Millman)

Figure 133. Basketball was also played at Portland, which can be seen in this photo. Notice the air raid shelter not far from the playing area, which quickly reminded those playing of why they were here. (Don Millman)

Figure 134. *PT-510* from Squadron 34 sits on the large pier in Portland. Standing on the pier with the long Navy trench coat is Lt. (jg.) Elliot B. Macswan, the boat's skipper. Portland was a massive base with many facilities, including a train, which would run supplies to the docks (notice the tracks in the dock) making it easier for the base force to do their jobs. (Don Millman)

Figure 135. The crew of *PT-502* (center) uses boat hooks to pull in closer to these other boats, as Ens. Buell T. Heminway maneuvers the boat. The other boats in the harbor are from Squadron 35. *PT-521* (far boat pier side) is commanded by Lt. Peter S. Zaley. One had to be careful as there was not much room for the boats with other British Coastal Forces craft to contend with. (Don Millman)

Figure 136. Moving day? *PT-507* has a large number of crates and chairs lashed to the foredeck. The tubular object is a handmade fender, used to protect the sides of the boat when moored to another boat or dock. The bases in England and France had wonderful facilities for the boats, which included stores, equipment and machine shops. Notice the different clothing worn by the crew. The sailor standing behind the 37-mm cannon, talking with the skipper and another sailor, is sporting a USAAF B-3 leather sheepskin lined bomber jacket. The rest of the crew is wearing a variety of foul weather jackets. Weather in this area was unpredictable, damp, and cold. (Don Millman)

Figure 137. A strange look for an American PT boat: a jeep lashed down on the bow. Cherbourg was opened as an advanced base in August 1944. This photo was snapped as the boat entered the harbor. (Don Millman)

Figure 138. Mugging for the camera is this Ron 35 crewmen. Notice he is sporting the latest in flak vests, usually reserved for bomber crews. These vests did not find favor among the PT crews because of their weight and sheer discomfort. The man behind him is observing the crew on the other boat who look like they are securing either a MK-X Mod 7 or a MK-XII naval mine to a launch rack. (Don Millman)

Figure 139. These Higgins PT boats are docked side by side in Cherbourg, France. They are from Squadron 30, which had action in the English Channel area from June 1944 to June 1945. This Squadron was one of the very few that had her boats shipped home for reconditioning. While the boats were still in New York waiting for reassignment to the Pacific, the war ended. (PT Boats Inc., Germantown, Tennessee)

Figure 140. A Squadron 34 boat maneuvers to moor next to this British Coastal Forces craft. Notice the US star that has been painted on her bow for aircraft recognition purposes. These were painted on the boats in hopes that Allied aircraft would spot the star from the air, and not accidently attack the boat. No incidences of this type attack have been recorded in the European Theater, although this was not the case in the Pacific. This boat also carries two 37-mm cannons, just forward of the torpedoes. (Don Millman)

Figure 141. Squadron 31 boats are docked at Teguchi Harbor, Okinawa. This photo was taken in June 1945, where the squadron was based until August. These Higgins PT boats had little to do here as targets for the boats had disappeared. They did, however, carry out successful air-sea rescue missions, and stood ready to disrupt any evacuations from Okinawa by enemy troops. In the course of these patrols they sank a lugger and six canoes, all loaded with troops trying to escape. Notice the Air-Sea Rescue boats docked opposite the PT boats. This would be known as base 24/25. (PT Boats Inc., Germantown, Tennessee)

Figure 142. Men of the Base Force pose in front of the large sign made up at the entrance to Base 17. This would become the largest PT boat base in the world, and was the last stop for many men before heading home at the end of the War. (Author's collection)

Figure 143. A nice look showing the massive main PT boat base in the Philippines. This was Base 17, located at Bobon Point in Samar. Seabees built and constructed many different sized Quonset Huts and buildings to accommodate the massive amount of men and supplies that made up this complex. Rains, typhoons, and frequent air raids interfered greatly with the servicing of the boats by the tenders, while the base was being constructed. The base was a sea of mud because of storms making it difficult to exist in the area. During the time of construction, the Floating Equipment –drydocks, radar barges, ramp barges, crane barges, gas barges, work shop barges, and the four F Ships and two FS Ships assigned to the task group were the saving features during early construction. (Author's collection)

Figure 144. Here we see the Officers Club at Base 17. After the war, this area was staffed by young women who were volunteers from the Red Cross. Notice the rather large bar area set up inside the club. The large sign at the bar is inviting officers to join in on the Regatta race. Dances and other activities were normal after the war. The woman in the picture is Darcy Kirk from Massachusetts who worked at Base 17 until it closed. (Darcy Kirk)

Figure 145. This is the main entertainment area (nicknamed the Bobon Bowl) at Base 17. This massive area has its own large movie screen (in the covered Quonset) and hand-crafted log benches for those in attendance. It was not uncommon to see this area crowded with officers and enlisted men each night watching a movie flown in from the USA. (Author's collection)

Figure 146. A look at the many dry docks brought over to Base 17. Boats from all over the Pacific pulled into this base for all types of repair. Most were then decommissioned. Here *PT-318* from Squadron 9 has just pulled into the dry dock for hull repairs. Several other boats can be seen in the harbor, making their way in. (Author's collection)

Figure 147. A torpedo carrier has been brought over to the dock to load this MK-XIII Torpedo. Supplies will be brought on to the boat to include ammo, food, water, and whatever else was needed to ready the boats for the night's mission. Men coming in from patrol would not rest until these chores were completed. Later the boat will be taken over to the fuel dock to top off her thirsty tanks. (Bob Hostetter)

Figure 148. *PT-336* fires up her engines during dock side trials. Base Force crews have just finished changing out two of her three Packard Marine Engines. Shortly after warm up, the boat will be run at speed to insure all is running smooth. At this time material supply to the South Pacific was the best of any time during the war and shortages of engines were rare. (Author's collection)

Figure 149. The main headquarters of the MTB Decommissioning Detail on Base 17. The planning for the disposal of the boats was carried out from this office, which was kept very busy with all of the details. To the right is the jeep driven by the commander of the base, J. Alex Michaud, who had worked for the Bureau of Ships prior to command of the base. Michaud made many training films for the Navy on the operation of the PT boat and served in both the North Atlantic and the South Pacific. (Author's collection)

Figure 150. A wonderful look at this eighty-foot Elco PT boat as she makes her way back to Base 17. Boats racing back during the daylight hours were in constant danger of enemy air attack, and sometimes were mistaken for enemy boats by our own Army Air Force. Several times, boats were attacked by our own fighters and bombers, with injuries and casualties on both sides. (Author's collection)

Figure 151. Getting the boats ready for decommissioning took plenty of hard work. Here, the ensign in charge of this work is checking the wiring harness for the center console. He is checking the connections for the RPM and manifold pressure gauges. Taken at Base 17 in Samar. (PT Boats Inc., Germantown, TN.)

Figure 152. Some of the boats still have their tent structures on the bow. These were set up so that gunner's mates could work on field stripping of the guns without worrying about the hot Pacific sun. The boats here are nearing the final phase of decommissioning, as their powerful Packard Marine Engines have been removed, as well as all armament and most essential equipment. (Randall J. McConnell)

Figure 153, A PT boat is not really at home until she is in the water. Here, *PT-589* is airborne as she is lowered into the bay at Samar. The boat served with Squadron 40. Base force personnel are seeing to the unloading of the boat. (Robert James Douglas)

Figure 154. *PT-144* is tied to the dock at Base 17. Base force personnel have already started to strip all essential parts. A list of all items that were to be removed and saved was handed down from the commander's office. The decommissioning detail went about their work of saving and scrapping items from each boat, with saved items crated for shipment home or sent to other destinations. This boat served with Squadron Two before being transferred to Squadron Eight. (Randall J. McConnell)

Figure 155. PT-146 would serve with three different squadrons before ending up here at Base-17, Bobon Point, Samar in the Philippines. These Base Force men are completing the task of stripping the boat of all useable equipment which would be stored and shipped to the United Sates. Much of the equipment taken off the boats such as guns, torpedoes, engines, radios, and other gear were packed in crates to be shipped. Although this was the procedure, much equipment was simply left in place or bulldozed into the jungle. (Robert James Douglas)

Figure 156. This unknown Higgins PT boat is crowded with base personnel as they continue to strip the boat of its equipment. Many times before the boats would be decommissioned, crews took them out for one last spin around the harbor, giving rides to those that who had never ridden on one of the boats before. Much of the equipment that was removed from these boats would never make it home, being deep-sixed into the ocean or simply left in place. Officers, enlisted men and base forces were only interested in getting home and Navy discipline became very relaxed and often over-looked. (Author's collection)

Figure 157. Boats are strung together at Base 17 after being stripped of all usable equipment. Wooden-hulled boats could not be mothballed as easily as steel ships, and no attempts were made to bring these "expendables" home. The three center boats are left to right: *PT-84, PT-222,* and *PT-221.* (Author's collection)

Figure 158. In this photo taken in the lagoon at Base 17, you can clearly see the helm area has been stripped of all gauges and other equipment. This boat is now ready for burning. Notice the holes that have been punched into the decking, which will be filled with gasoline, once the boat has been towed away from the dock. (Alvin Hansen)

Figure 159. The engineering shop at Base 17 (center Quonset hut) shows how many 4M-2500 Packard Marine Engines were stripped from the PT boats after the war. These engines were stacked in this area with the intention of shipping them to the United States. In the end, most of the engines were simply bulldozed into the jungle, with some given to the Army to supply their boats, and others being dumped into the ocean. It was thought that because the Navy was not going to be sending the PT boats home (due to their wooden construction) there would be little need for stateside spares. (Lt. Everett L. Carrier)

Figure 160. Boats lined up at Base 17. These three center boats (from left to right) *PT-352, PT-114, and PT-120* have undergone major stripping of parts and armament. Notice the Packard engine exhaust stacks that are laid on the decks. (Alvin Hansen)

Figure 161. This photo shows all that is left of an eighty-foot Elco PT boat. Taken from the dock at Base 17, the boat has been stripped of all essential materials, filled with gasoline, and burned down to the waterline. By this time many of the crews from these boats had been rotated home, so it was the decommissioning staff that watched the boats' final hours. (Alvin Hansen)

Repair Training Unit

From the early war years it became evident that specialized repair instruction was needed rather than the jack-of-all-trades approach that was first implemented. This theory had been adopted because it was thought that on so small a craft, everyone should be able to pinch hit for each other. But this method failed to provide thoroughness and handicapped the forces in operating areas. Many of the repair problems of the first squadrons could have been avoided if the Repair Training Unit had been formed earlier.

On February 12, 1944, this unit was activated at the Motor Torpedo Boat Squadrons Training Center and became known as the Repair Training Unit or MTBRTU. It was commissioned on March 7th with a simple ceremony, satisfying a long-recognized need to provide experienced base forces for tenders and advanced bases. Personnel trained in maintenance, repair, and upkeep could affect great economy of materials in the combat area. The emphasis was on the practical rather than the theoretical. The Repair Training Unit maintained its own faculty who taught a curriculum pointed at maintenance rather than operations.

Trainees included officers and men, and those not assigned to AGP's were formed in E-11 and E-12 units. These units were advanced base functional components, and were used in operations where the immediate establishment of a large base was not feasible. The courses, in fact the whole idea of a RTU, indicated a trend toward higher standards. Students working alongside existing repair crews did the physical work by actually handling the tools on the job and gained experience of great value for their future stations. In six weeks personnel were to become as proficient as possible in a definite branch of repair. Some general information was covered: however each major category received detailed attention. These were engineering, electrical work, carpentry, carburetion, shipfitting, gunnery, and torpedoes. After this training, these men formed the repair gangs on tenders and advanced bases. In addition, a portion of the course was devoted to teaching useful information on establishing and maintaining bases and equipment under combat conditions with regard to climate.

As part of the duties as CO of the Motor Torpedo Boat Squadron Training Center, the unit was first under Commander D.J. Walsh, USNR. The Executive Officer was Lt. Mark E. Wertz who chose an experienced staff with Lt. E. L. Carrier as Senior Instructor.

All of the men who served under these instructors were experienced veterans who had served in the different theaters and brought back valuable lessons learned in the field. Since it was not practical for students to train entirely on existing equipment, shops were set up for their use. The student personnel had several years experience as skilled workers and this was carried over to their new rates whenever possible. In 1945 the program began to include cargo handling taught with models.

Lt. Robert A. Williamson succeeded Wertz in September 1945. Wertz received the Navy Commendation Ribbon and citation in recognition of his work in training repair crews.

Figure 162. Commander David J. Walsh, USNR was born on May 24, 1905 in Ansonia, Connecticut. He entered the United States Naval Academy and graduate with the class of 1927. He served on the *USS Wyoming, USS Rochester,* and *USS Galveston*, and left active duty in September 1929. He maintained a reserve commission, first working in the Engineering Department of the Western Electric Company with the Electric Boat Company in Groton, Connecticut. He then served as a naval engineer and inspector for the Office of the Inspector of Naval Material, Schenectady, New York. Recalled to active duty in January, 1941, then Lt. (jg) Walsh reported for duty onboard the first PT boat tender, *USS Niagara*, as the Executive Officer. He assumed command of the Motor Torpedo Boat Squadrons Training Center on September 18, 1943. As part of his duties as Commander of the MTBSTC, Walsh (in the beginning) was in charge of the Repair Training Unit. (Author's collection)

Figure 163. Lt. Marcus E. Wertz was the Executive Officer of the Repair Training Unit at the Motor Torpedo Boat Squadrons Training Center. Some officers and enlisted men returning from active duty were assigned to the Repair Unit (temporary) to pass on their experience to students. Wertz took over as Commanding Officer from Cmdr. David J. Walsh. Wertz received the Navy Commendation Ribbon and Citation for his work in training repair crews. Prior to joining the RTU, Wertz was the CO on *PT-14.* He also rode on *PT-28, 37, and 47,* serving with Squadrons One and Two. After the war, he would become President and owner of Gulfport Steel Fabricators. (PT Boats Inc., Germantown, Tennessee)

Figure 164. Lt. Everett L. Carrier (left) shown here in the cockpit of this Elco PT boat was the Senior Instructor of the Repair Training Unit. Prior to his orders to report to the Motor Torpedo Boat Squadrons Training Center, Carrier served at Tulagi in the Solomon Islands. While at Guadalcanal, he worked out deals with the B-17 crews that operated on the airfield there. He traded for fuel pumps, selector valves, and fuel lines needed to keep the boats running. He later become CO of Base 17 at Bobon, Samar in the Philippines. This was the largest PT boat Base ever constructed. (Lt. Everett L. Carrier)

Figure 165. Hands-on training for the Repair Training Unit students was vital and proved highly effective in the combat areas. This student is being watched carefully by his instructor as work begins on repairing the bow section of this PT boat. Although the boats were built to withstand the twisting and turning of the ocean, boats hit in combat by enemy fire needed repair to their wooden hulls and superstructures to continue operations. (PT Boats Inc., Germantown, Tennessee)

Figure 166. Repair Training Unit students learn first-hand the inner workings of the Browning .50 caliber machine gun, which was the main weapon on a PT boat. This set up is complete with turret ring and back rest. This is a working weapon on the firing range at the Motor Torpedo Boat Squadron's Training Center. (Author's collection)

Figure 167. RTU students are learning first-hand the inner workings of the Bofors 40-mm single-mount cannon. Here two students are working on the breech mechanism of the weapon while other students watch. This type of training was essential in the warzone and greatly benefited the overall repair program. (Authors collection)

Figure 168. Students at the base are loading gasoline fuel tanks onto a flat car with the help of a small crane. These are self-sealing center tanks and will be used to teach students the proper method of installation on the boats. Notice the 4M-2500 Packard Marine Engine (on dolly) to the right of the photo. The Quonset huts are used for storage and maintenance along the main concrete pier. (Author's collection)

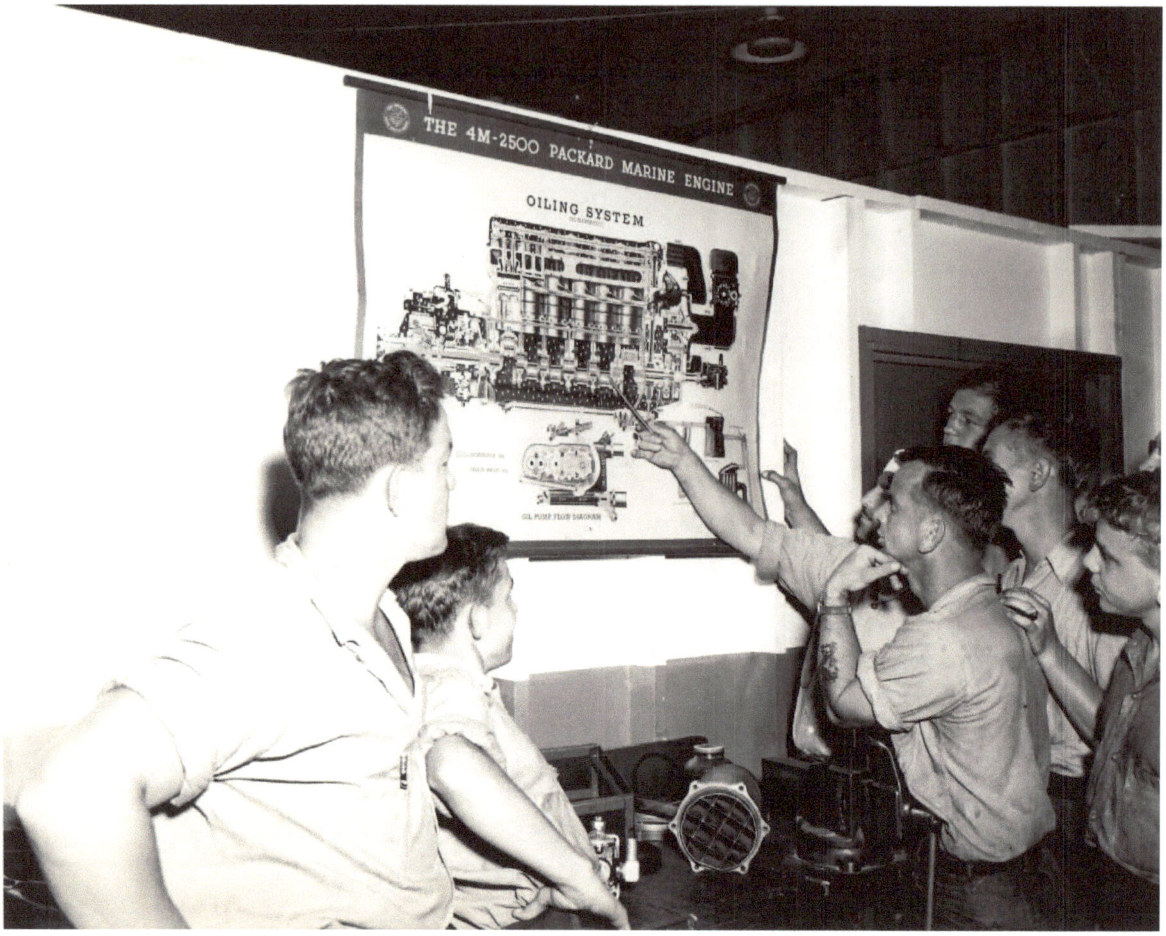

Figure 169. Students from the Repair Training Unit are learning about the oiling system of the 4M-2500 Packard Marine Engine. Classes of this nature would become instrumental when students would be sent to the War zone working hand and hand with the boats Motor Machinist Mates. Tearing down engines to replenish worn out parts would be crucial to the boats survival, as these high powered engines demanded fine tuning. (PT Boats Inc., Germantown, Tennessee)

Figure 170. Members of the Repair Training Unit at the Motor Torpedo Boat Squadron Training Center are removing this cabin trunk structure from this seventy-seven-foot Elco PT boat. Students worked side by side with existing repair crews so that they actually handled the tools on the job and gained experience of great value for their future stations. (Author's collection)

Figure 171. The dry docks near the finger piers of the Motor Torpedo Boat Squadrons Training Center. It was here that the Repair Training Unit learned first-hand the inner workings of the dry docks and proper placement and location of the boats. This Huckins boat *(PT-101)* has entered the dry dock for hull repairs. Also notice to the right a ten-ton crane on a floating pontoon barge, and to the left, a Quonset hut. Things became very cold at the base during winter as ice formed in the basin. This made for uncomfortable conditions for base force personnel, who were only too happy to be transferred to the warmer climate of Florida or New Orleans. (Author's collection)

Figure 172. A cannibalized Elco seventy-seven foot PT boat sits on her cradle at the Motor Torpedo Boat Squadrons Training Center. Repair Training Unit classes were often held in these areas to teach the inner workings of the boats in such areas as hull repair, engine repair, electrical work, and gunnery. Notice the athletic field that was part of the base (upper right of photo). (Author's collection)

Figure 173. RTU students work closely with their instructor. This boat has had its high- pitched props taken off, and students are checking the shafts, struts, and keyways. Boats in combat were in constant danger of damage to this area due to uncharted areas, rocks, or debris in the water. Repair crews needed to fix this type of damage in short order, if the boats were to remain on patrol. Training in this area was extensive. (Author's collection)

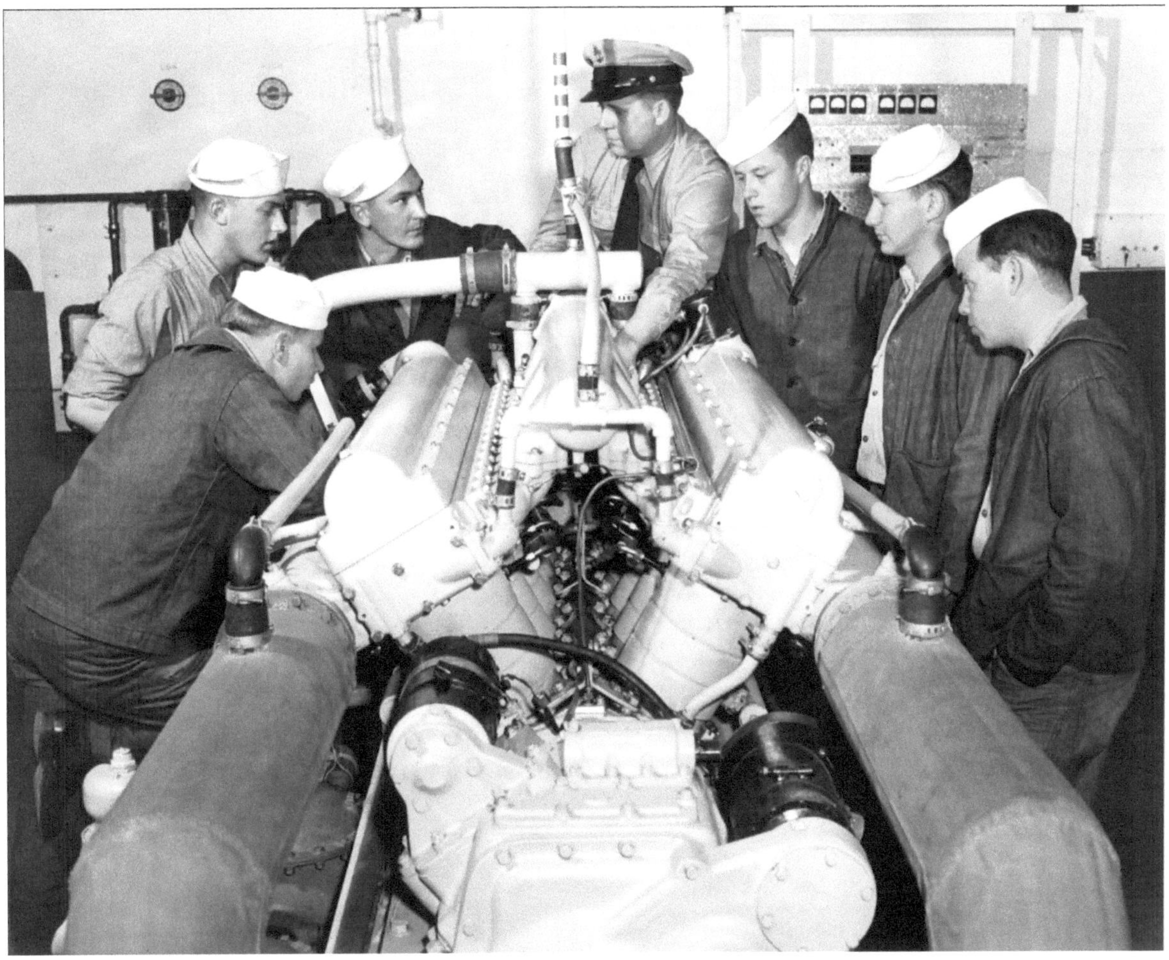

Figure 174. RTU students learn first-hand the inner workings of the 4M-2500 Packard Marine Engine. The instructor (standing in front of the supercharged end of the engine) is showing the students the carburetion system, ignition over-speed cutout, and supercharger. Notice the exhaust manifolds are piped to the outside, which means this is a working engine. (Author's collection)

Figure 175. Students of the Repair Training Unit were trained with the belief that hands-on training would benefit them in the forward combat areas. This led to much more efficient maintenance program. Here two students are deep inside the engine room of this PT boat, checking on the exhaust stacks and performing routine maintenance. (Author's collection)

Figure 176. RTU students are busy learning about the reverse gear end of the engine. The 4M-2500 Packard Marine Engine contained its own built-in reverse gear. This gear transmitted the power of the engine to the propeller, providing both forward and reverse drives, as well as a neutral gear to allow the engine to run idle. These visual aids were a constant reminder to the students and enhanced the lessons learned. (PT Boats Inc., Germantown, Tennessee)

Figure 177. Students are learning the proper way of securing and loading torpedoes. With the help of the torpedo tug, the torpedo is now in place and the securing straps are being removed. RTU students at the base were instructed by officers who had recently come back from the war zone and could provide the trainees with first-hand knowledge. (Authors collection)

Figure 178. A look at the engine shop located at the base. Students were taught how to overhaul the 4M-2500 Packard Marine Engines. Because the engine was not your typical slow-speed heavy duty marine engine, it required the same expert handling in its operation, maintenance, and overhauling an aircraft's power plant. (Author's collection)

Figure 179. Students gather around the main concrete dock. Here, instructors will teach them the complete workings of the dry dock system (pictured here). Students are learning how to run the controls to flood and counter flood the lift. Notice the cradle for the boats has started to lift out of the water. The PT boats were raised to a position in which base force and crew could work comfortably on the bottom of the hull. (PT Boats Inc., Germantown, Tennessee)

Figure 180. A look at the electrical shop located on the base. Base force electricians needed to know the complete electrical system of the PT boats, including radios, radar, engine room panels, cockpit panels, and wiring. This shop tested many of the components that make up the boat. (PT Boats Inc., Germantown, Tennessee)

Figure 181. This instructor is teaching these students about the MK. 50, five-inch spin-stabilized rocket launcher and MK. 8 lluminating sight (for fixed gunnery firing). The instructor will demonstrate bore sighting, loading and non loading procedures, safety precautions, and harmonization of sight launchers. The additions of these launchers gave the little PT boat the punch of a destroyer, although they were added too late in the war and were used very little in overall operations. (PT Boats Inc., Germantown, Tennessee)

Figure 182. Graduating class members from the Repair Training Unit line up in front of their building, at the Motor Torpedo Boat Squadrons Training Center. They have completed the six week course in a designated field, which included engineering, electrical work, carburetion, ship fitting, gunnery, and torpedoes. These man would form the repair gangs on tender and advanced bases. (PT Boats Inc., Germantown, Tn.)

Figure 183. Repair Training Unit Officers and some of their Students take the time to pose in front of the RTU training facilities at the Motor Torpedo Boats Squadrons Training Center, located at Melville. (Author's collection)

FEMU Units

The Floating Equipment Maintenance Unit was a conglomeration of people and barges. These were pontoon barges with a Quonset hut in the middle, a ten-ton crane on the bow, and a couple of pusher units, using eight cylinder Chrysler marine engines aft. At full rpm and no tide, two knots could be attained. These floating dry docks were used to pick PT boats out of the water, in order to dry them out or for crews to do necessary repairs below the water line. The barges could change out torpedoes, or Packard Marine Engines, as well as provide limited supplies to the boat crews. These barges could operate in the forward areas, getting as close to shore as possible, or anchoring out in the harbor. Their worth was that they could follow the squadrons in the forward areas providing maintenance. There was an assortment of fuel barges, carrying from 80,000 to 100,000 gallons of high-octane gasoline to fuel the boats. This was certainly a dangerous job, as one well placed hit from enemy guns or planes would destroy anything in a large area. FEMU had roughly 275 officers and men at any given time. The unit was formed at Base 21 in Mios Woendi, New Guinea in 1944, and was moved to a small island opposite Base 17, in the Surigao Strait near Tacloban, Philippines. In 1945, it was moved with Base 17 to Samar, in the Philippines. The barges and attending personnel traveled all over the Philippines', either to their bases or with their tenders, escorting squadrons on their various assignments. After the war ended, the barges, just like the PT boats were scrapped and the personnel reassigned or sent home.

Figure 184. A look at most of the Officers and enlisted men, that made up the Floating Equipment Maintenance Unit. They were well trained in providing the boats with much-needed tender Barges, Floating Dry Docks, Gasoline Fuel Barges, and other types of floating equipment. These barges enabled the PT Squadrons to move feely wherever the enemy was located. These men were well trained and well accepted by the PT boat crews. (Alfred B. Moore)

Figure 185. A look at PT-47 being off-loaded at Noumea, New Caledonia. This wonderful crane set up appears to be one of the Seabees heavy lift crane barges. It was built using the famous pontoon sections (look close at the deck of the crane barge). The design of the crane was made for the off-loading of small craft from transports. PT-47 would serve with Squadron 2 and Squadron 3(2), before being shipped to the MTBSTC (Randy Finfrock)

Figure 186. A nice look at Crane Barge #6 attached to the Floating Equipment Maintenance Unit. Notice the large ten-ton crane on her deck, used to lift everything from Torpedoes, and Engines to the waiting PT boats. If you look closely you can see one of the center fuel tanks lying next to the Quonset hut. The huts were used as sleeping quarters for at least four men, as most equipment was removed and used on the beach. There was a ping-pong table inside as well as a dark room used by Al Moore, who developed photos. (Alfred B. Moore)

Figure 187. The stern end of Crane Barge #6, showing her ten-ton crane set up. She is tending to a Squadron 24 PT boat. Many times boats returning from patrol would simply look for dock space, tying up to the FEMU Barges. Barges were made by the Seabees in the War Zone from designs by the Navy, using two eight cylinder Chrysler marine Pushers, giving them four knots with no tide. The four man crew on this barge consisted of Gerald K. Barnett, Machinist Mate 1/c, Edward S. Miller, Bosons Mate 2/c, Alfred B. Moore, Motor machinist Mate 3/c, and Harold Anderson, Fireman. (Alfred B. Moore)

Figure 188. Alfred B. Moore, running the ten-ton crane on crane barge #6. He has hooked up one of the MK-XIII Torpedoes and will very carefully lower it to the waiting roll-off rack of a Squadron 24 boat. On March 9, 1945, a tug towing this barge, two fuel barges and a dry dock strung out in single file a couple of hundred yards apart, headed for Zamboanga. Halfway down the coast of Mindanao tow lines parted between the two gasoline barges and the dry dock. This happened three times and on each occasion the accompanying tug circled the barges looking for a way to get a line over to the loose barges. Alfred B. Moore and Edward S. Miller jumped into the water, each time securing temporary tow lines. For their actions in trying to secure the barges, both men were awarded the Navy Marine Corps Medal. (Alfred B. Moore)

Figure 189. A look at the dry dock area in Morotai on Halmahera Island shows these PT boats being attended to. *PT-126* (in the water) carries the latest in armament with rocket launchers, 37-mm auto-cannon, 40-mm Bofors cannon, and the MK-XIII Torpedoes in roll-off racks. This boat served with Squadron 6. The other boat (already in dry dock) is *PT-159,* which served with Squadron 9. These floating dry docks were crucial in the forward areas to insure the boats received the utmost in maintenance. (Randall J. McConnell)

Figure 190. FEMU Crane operator steadies his boom as the platform for this boat's 40-mm Bofors is being removed for service. The heavy ten-ton crane was essential in the removal of equipment too heavy for men to lift. (Authors collection)

Figure 191. A barge from the FEMU tends to these two Elco PT boats. On the left is *PT-330* from Squadron 21, and on the right, *PT-160* from Squadron 9. Notice the grass thatch roof on the barge, which is also hauling a dry dock unit to the rear. It does not appear that this barge has the typical ten-ton crane as part of her equipment. These barges provided a place for the boats to dock, supplies, food, fuel, dry docks for hull repairs, and cranes for engine replacement and torpedo changing. (PT Boats Inc., Germantown, Tennessee)

Figure 192. Dry docks on the advanced base on the Morobe River in New Guinea were a welcome addition. The docks were towed to their destination and provided base forces a dry place for hull repair. They were narrow basins that were flooded to allow the PT boat to be floated in, and then drained to allow the boat to come to rest on a dry platform. Base force and boat crews then performed a multitude of tasks, such as shaft, strut and propeller repair. Boat bottoms were scraped, painted and, most importantly, dried out. (Gene Kirkland)

Figure 193. This crane barge has hooked onto one of the MK-XIII torpedoes. Notice the Quonset hut used to store supplies and provide sleeping quarters for the FEMU crew. Some of the larger huts provided small areas for machine repair (limited) and could provide food for PT boats that docked on the barge. (Alfred B. Moore)

Figure 194. PT boats docked at advance base 4, Halmahera Island, in June 1945. To the left of the photo you will notice a ten-ton crane at work, with one boat in dry dock. This was part of the Floating Equipment Maintenance Unit. To the center of the photo, *PTs 177, 176,* and *175* from Squadron 11 are strung together. This base offered little protection from shifting winds and seas. During the time spent here, it was a constant struggle to keep the boats and floating equipment from blowing aground on reefs. It was an excellent location however to intercept Japanese barge traffic between Manokwari and Sorong, the principle enemy base on the Western end of Vogelkop. It is from this base that one of the largest enemy vessels was sunk in the New Guinea campaign. *PT-342* torpedoed a 200-foot ship thought to be a minelayer. (National Archives)

Figure 195. This FEMU crew is being kept busy as they work on two boats from Squadron 34 in England. Notice the boat in the background has been put into the floating dry dock, while the boat in the front is in the water. This FEMU barge is equipped with a ten-ton crane, generator, floating dry dock, and service dock. (PT Boats Inc., Germantown, Tennessee)

Figure 196. *PT-261* from Squadron 26 has been prepared and put into dry dock. This boat was a Huckins-built design, and served with the Hawaiian Sea Frontier. This Squadron spent its time in training and in patrolling coordinated areas, but saw no action with the enemy. This boat was undergoing hull scraping and painting when this photograph was taken. (Lt. Cmdr. Lester H. Gamble)

Figure 197. These Navy Seabees are hard at work as they finish construction on this fuel barge. These barges were also part of a FEMU and proved to be very dangerous duty. Containing hundreds of gallons of high-octane fuel, and towed, they were highly vulnerable to air attack. This photo was taken at Mios Woendi, New Guinea. (National Archives)

Figure 198. This FEMU barge has managed to nose itself close to the beach. Taken at the temporary base on Amsterdam Island, Dutch New Guinea, the barge already has a customer (just beginning to dock), and will do whatever is needed for support of the boat. Notice to the right of the barge, the towed fuel barge with its two large high-octane fuel tanks. (National Archives)

Figure 199. These Squadron 24 men are unloading camp gear supplies on Amsterdam. Notice how close the two PT boats are to the beach. This base would prove difficult due to shifting winds and seas, causing constant problems for the barges, which could be blown onto reefs. (National Archives)

Figure 200. USS *Mobjack* (AGP-7), an *Oyster Bay*-class tender. (Robert E. Pickett)

www.ingramcontent.com/pod-product-compliance
Ingram Content Group UK Ltd.
Pitfield, Milton Keynes, MK11 3LW, UK
UKHW061541051026
15625UKWH00011B/562
* 9 7 8 1 6 0 8 8 8 1 4 9 9 *